Thin Film Transistor

Thin Film Transistor

Special Issue Editor

Ray-Hua Horng

MDPI • Basel • Beijing • Wuhan • Barcelona • Belgrade

Special Issue Editor
Ray-Hua Horng
Institute of Electronics, National Chiao Tung University
Taiwan

Editorial Office
MDPI
St. Alban-Anlage 66
4052 Basel, Switzerland

This is a reprint of articles from the Special Issue published online in the open access journal *Crystals* (ISSN 2073-4352) from 2018 to 2019 (available at: https://www.mdpi.com/journal/crystals/special_issues/Film_Transistor)

For citation purposes, cite each article independently as indicated on the article page online and as indicated below:

LastName, A.A.; LastName, B.B.; LastName, C.C. Article Title. *Journal Name* **Year**, *Article Number*, Page Range.

ISBN 978-3-03921-526-3 (Pbk)
ISBN 978-3-03921-527-0 (PDF)

© 2019 by the authors. Articles in this book are Open Access and distributed under the Creative Commons Attribution (CC BY) license, which allows users to download, copy and build upon published articles, as long as the author and publisher are properly credited, which ensures maximum dissemination and a wider impact of our publications.

The book as a whole is distributed by MDPI under the terms and conditions of the Creative Commons license CC BY-NC-ND.

Contents

About the Special Issue Editor . . . . **vii**

Fu-Ming Tzu and Jung-Hua Chou
Effectiveness of Light Source on Detecting Thin Film Transistor
Reprinted from: *crystals* **2018**, *8*, 394, doi:10.3390/cryst8100394 . . . . **1**

Fu-Ming Tzu and Jung-Hua Chou
Optical Detection of Green Emission for Non-Uniformity Film in Flat Panel Displays
Reprinted from: *crystals* **2018**, *8*, 421, doi:10.3390/cryst8110421 . . . . **9**

Ray-Hua Horng, Ming-Chun Tseng and Dong-Sing Wuu
Surface Treatments on the Characteristics of Metal–Oxide Semiconductor Capacitors
Reprinted from: *crystals* **2019**, *9*, 1, doi:10.3390/cryst9010001 . . . . **22**

Hee Yeon Noh, Joonwoo Kim, June-Seo Kim, Myoung-Jae Lee and Hyeon-Jun Lee
Role of Hydrogen in Active Layer of Oxide- Semiconductor-Based Thin Film Transistors
Reprinted from: *crystals* **2019**, *9*, 75, doi:10.3390/cryst9020075 . . . . **31**

Markus Krammer, James W. Borchert, Andreas Petritz, Esther Karner-Petritz, Gerburg Schider, Barbara Stadlober, Hagen Klauk and Karin Zojer
Critical Evaluation of Organic Thin-Film Transistor Models
Reprinted from: *crystals* **2019**, *9*, 85, doi:10.3390/cryst9020085 . . . . **38**

Jui-Fen Chang, Hua-Shiuan Shie, Yaw-Wen Yang and Chia-Hsin Wang
Study on Correlation between Structural and Electronic Properties of Fluorinated Oligothiophenes Transistors by Controlling Film Thickness
Reprinted from: *crystals* **2019**, *9*, 144, doi:10.3390/cryst9030144 . . . . **56**

Jiung Jang, Yeonsu Kang, Danyoung Cha, Junyoung Bae and Sungsik Lee
Thin-Film Optical Devices Based on Transparent Conducting Oxides: Physical Mechanisms and Applications
Reprinted from: *crystals* **2019**, *9*, 192, doi:10.3390/cryst9040192 . . . . **70**

August Arnal, Carme Martínez-Domingo, Simon Ogier, Lluís Terés and Eloi Ramon
Monotype Organic Dual Threshold Voltage Using Different OTFT Geometries
Reprinted from: *crystals* **2019**, *9*, 333, doi:10.3390/cryst9070333 . . . . **81**

Ray-Hua Horng
Thin Film Transistor
Reprinted from: *crystals* **2019**, *9*, 415, doi:10.3390/cryst9080415 . . . . **96**

About the Special Issue Editor

Ray-Hua Horng received her B.Sc., and Ph.D. degrees from National Cheng Kung University and National Sun Yat-Sen University, Taiwan, in 1987 and 1993, respectively, all in electrical engineering. She has worked in the field of III–V compound materials by MOCVD and as Associate Researcher with Telecommunication Laboratories, Chunghwa Telecom Co. Ltd., Taoyuan, Taiwan. She is currently Distinguished Professor with the Institute of Electronics, National Chiao Tung University. She has authored or coauthored over 300 technical papers and holds over 100 patents in her fields of expertise. Her main interests are solid-state lighting devices, III–V solar cells, optoelectronic devices, high power devices, nanosurface treatment by natural lithography, power devices, and gas sensors. In 2000, she vitalized her research on high-brightness LEDs with mirror substrates into practical mass-produced items that enable high-power LEDs. Dr. Horng has received numerous awards recognizing her work on high-brightness LEDs. She has been awarded by the Ministry of Education of Taiwan for Industry/University Corporation Project in 2002; by the Ministry of Science & Technology of Taiwan for the excellent technology transfer of high-power LEDs in 2006, 2008, 2009, 2010, and 2011; by Chi Mei Optoelectronics for the first prize of Chi Mei Award in 2008; by the 2007 IEEE Region 10 Academia–Industry Partnership Award; and received the Distinguished Research Award of National Science Council of Taiwan in 2013. She has been Fellow of the Australian Institute of Energy since 2012, Fellow of the Institution of Engineering and Technology since 2013, Fellow of SPIE since 2014, Fellow of IEEE since 2015, and Fellow of OSA since 2016.

Article

Effectiveness of Light Source on Detecting Thin Film Transistor

Fu-Ming Tzu [1,*] and Jung-Hua Chou [2]

1. Department of Marine Engineering, National Kaohsiung University of Science and Technology, Kaohsiung 80543, Taiwan
2. Department of Engineering Science, National Cheng Kung University, Tainan 70101, Taiwan; jungchou@mail.ncku.edu.tw
* Correspondence: fuming88@nkust.edu.tw; Tel.: +886-7-810-0888 (ext. 25245)

Received: 9 September 2018; Accepted: 19 October 2018; Published: 21 October 2018

Abstract: Light sources tend to affect images captured in any automatic optical inspection (AOI) system. In this study, the effectiveness of metal-halide lamps, quartz-halogen lamps, and LEDs as the light sources in AOI systems for the detection of the third and fourth layers electrodes of thin-film-transistor liquid crystal displays (TFT-LCDs) is examined experimentally. The results show that the performance of LEDs is generally comparable or better than that of metal-halide and quartz-halogen lamps. The best optical performance is by the blue LED due to its spectrum compatibility with the time-delay-integration charged-coupled device (TDI CCD) sensor and its better spatial resolution. The images revealed by the blue LED are sharper and more distinctive. Since current LEDs are more energy efficient and environmentally friendly, using LEDs as the light source for AOI is very beneficial. As the blue LED performs the best, it should be adopted for AOI using TDI CCD sensors.

Keywords: metal-halide lamp; quartz-halogen lamp; blue LED; TFT-LCD; spectrum

---

1. Introduction

In the market of flat panel displays, especially the larger sizes, thin-film-transistor liquid crystal displays (TFT-LCDs) are currently the dominant product. With the progress in manufacturing, the product is moving from the high-definition television (1920 × 1080 pixels) of about 6 million subpixels toward the ultra-high resolution television of 10 million pixels and beyond. The latter has wide view illumination, sharp contrast, fast response, lower power consumption, and minimum radiation [1,2]. The wide view illumination coupled to the ultra-high definition (UHD) of 4K (3840*2160) is expected to move to 8K (7680*4320), 16K (15,360*8640), and 32K (30,720*17,280) [3] as the technology advances. The TFT of LCDs investigated in this study is fabricated by the back-channel etching process. It consists of five layers as (1) gate metal, (2) TFT layer (gate dielectric/channel/n+), (3) source/drain (S/D) metal, (4) silicon nitride (SiNx) passivation layer, and (5) indium tin oxide (ITO) pixel electrode. Additionally, there is low resistance gate metallisation using aluminium or copper, capped by chromium (Cr) on the third layer [4–6]. They are typically deposited either by physical vapour deposition (PVD) or by plasma enhanced chemical vapour deposition (PECVD).

Among these five layers, the third and fourth layers control both the light switching function of the liquid crystal and the frame rate of the LCD. Thus, in this study, the quality of these two layers is examined by an in-line automatic optical inspection (AOI) system for which the light source plays a key part. Presently in the display industry, the main light sources for AOI are the metal-halide and quartz-halogen lamps. Metal-halide lamps generally have a lifespan values range from of 6000 to 15,000 h [7,8] and provide good colour rendering due to their high-intensity discharge (HID) characteristic. However, their functioning requires 250 W_p. They are also very sensitive to voltage levels. If the operation voltage was lower than 220 V, the output light will decay immediately and may even shutdown completely. Moreover, they need warmup times (averaging a couple minutes) for stable operation. In addition, the lighting intensity tends to vary from lamp to lamp. In contrast, quartz-halogen lamps radiate significant amounts of heat with a lifespan of about 2000 h and cost more. Furthermore, the halogen elements are harmful to both human health and to the environment, and do not conform to the Waste Electrical and Electronic Equipment Directive (WEEE) and Restriction of Hazardous Substances [9]. Hence, with the growing concerns of global warming and its impact on the environment, a light source that is environmentally friendly and offers energy savings is of interest, and light emitting diodes (LEDs) are a potential alternative.

LEDs are solid state semiconductor devices of p-n junction diodes. They are highly energy efficient with an attainable lumen per watt of ~200 (lm/W), which is much better than both HID and halogen lamps. Furthermore, they contain no halogen elements. Namely, they are both energy efficient and environmentally friendly. Chulkov et al. [10] applied both LEDs and halogen light sources to inspect metallic materials which could lead to corrosion by active thermal waves. The effect of paint-and-lacquer coating colour on the heating efficiency using these light sources was analysed. The possibility of using LED thermal stimulation in portable flaw detectors was then described. The results showed that the LED performs well and is cost effective and suitable for AOI. Thus, LEDs look promising for AOI applications. Hence, in this study, the suitability of using LEDs in AOI of TFTs is examined by comparing their performance with commonly adopted light sources in the TFT-LCD industry.

2. System Architecture

The experiments were conducted in a class 1000 clean-room at 25 °C using tailor-made samples of the 6th generation glass panel with TFT electrode pixels. For the inspection of the third and fourth layers of TFT structures by image scanning, a line-scan of time-delay-integration (TDI) of a charge-coupled device (CCD) was employed. The TDI CCD can capture more images with the pixels in synchronization of the moving object, thus allowing the data packet to continuously track the motion of the object [11,12].

A commercial off-the-shelf HS 8 K TDI CCD (Piranha HS 8 K 68 kHz, TELEDYNE DALSA) was adopted for this task. Its photoelectric sensors can scan the images in hundreds of thousands of lines per second at very high speed and can also operate under low light levels and slower speed conditions if necessary. The optical resolution of the sensor is 1 µm; the wavelength is from ultra-ultraviolet (UV) to infrared (IR) with the maximum quantum efficiency of 38% occurring at the wavelength of ~520 nm. In other words, this device captures multiple exposures of the moving object to achieve higher responsivity. Figure 1 depicts a schematic diagram of the measurement setup. A computer-controlled gantry was installed to scan the samples that were held with a non-reflective film to avoid light interference. The main components of the scan model include the TDI CCD, the light source, a focusing lens, a fiber, a spectrometer, and a host computer. Reflected lights from the sample were captured by the CCD and fed to the automatic data acquisition program in the host computer for data analysis.

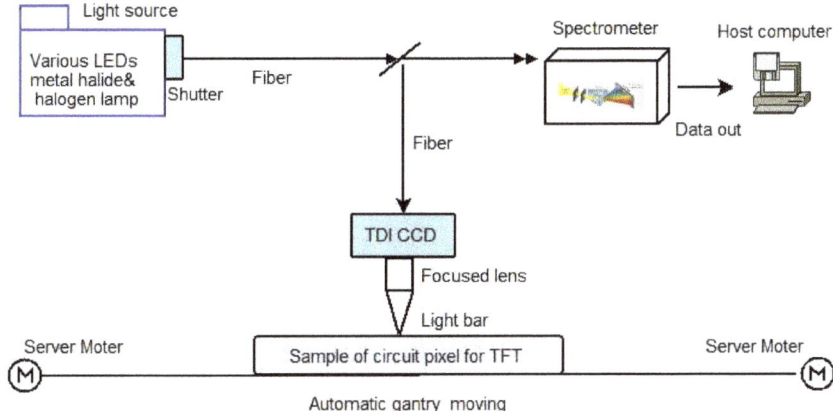

Figure 1. Scanning mode of AOI architecture.

For the purpose of comparison, the light sources employed included a 250 W_p metal-halide lamp (PCS-UMX250, COLDSPOT, NPI), a 250 W_p quartz-halogen lamp (MHF-KFB100LR, MORITEX) and monochromatic LEDs. Since the light spectra of metal-halide and quartz-halogen lamps are wide band (described in the next section), red, orange, yellow, green, and blue optical filters (see Figures 2 and 3) were used to narrow the light spectrum range.

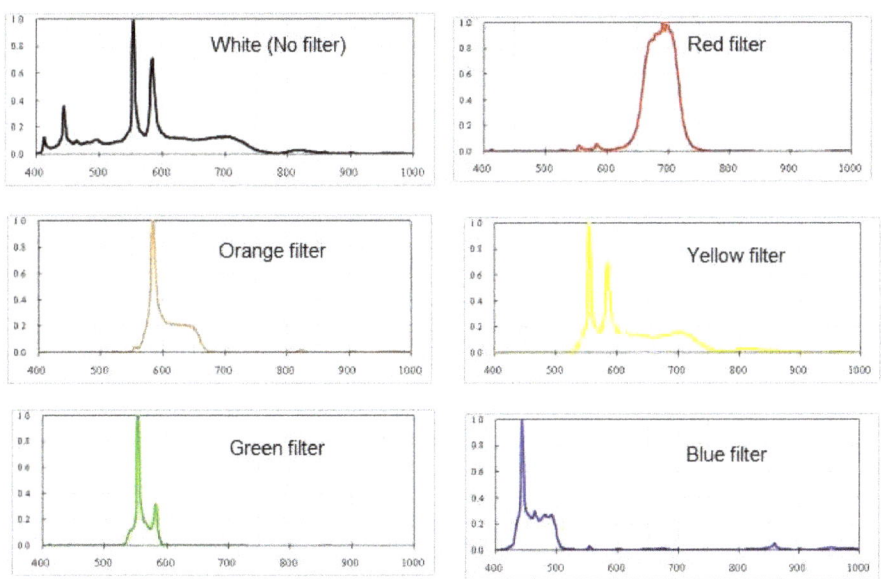

Figure 2. Spectrum distribution of metal-halide light source.

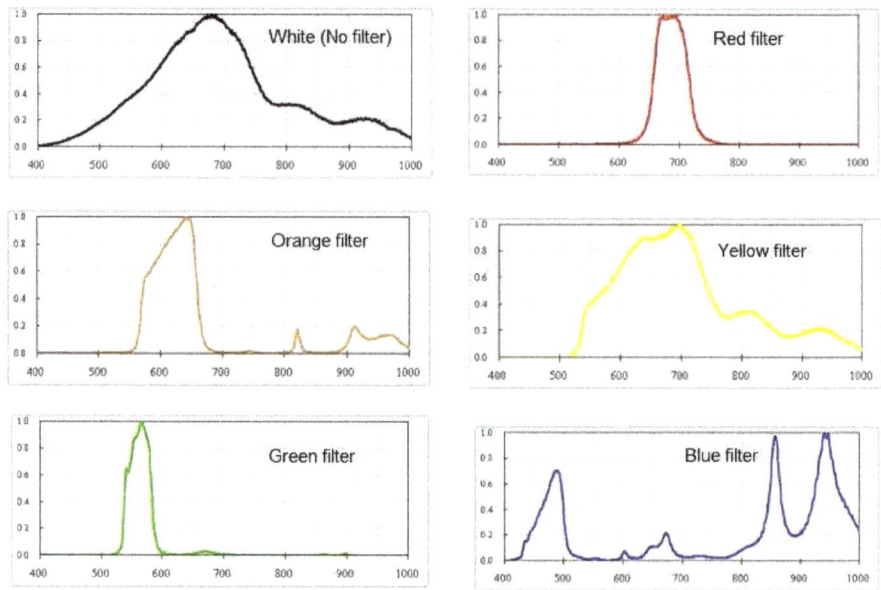

Figure 3. Spectrum distributions of quartz-halogen light sources.

The characteristics of these filters are tabulated in Table 1 for reference. In contrast, as shown in Figure 4, the spectra of the applied LEDs were monochromatic, that is, red, green, and blue with wavelengths of 640 nm, 525 nm, and 460 nm, respectively; no filters were needed. They are InGaN-based high-brightness LEDs, driven by a forward maximum current of 27 A and provide 5500 lumens at 90 W_p.

Table 1. Characteristics of the optical filters in the light source.

Filter	Type	Wavelength	Tran.$_{avg}$ (%)
Band-pass	Red	687 ± 30 nm	91%
Band-pass	Orange	615 ± 45 nm	94%
Longwave-pass	Yellow	≥555 nm	90%
Band-pass	Green	559 ± 22 nm	89%
Band-pass	Blue	465 ± 33 nm	85%

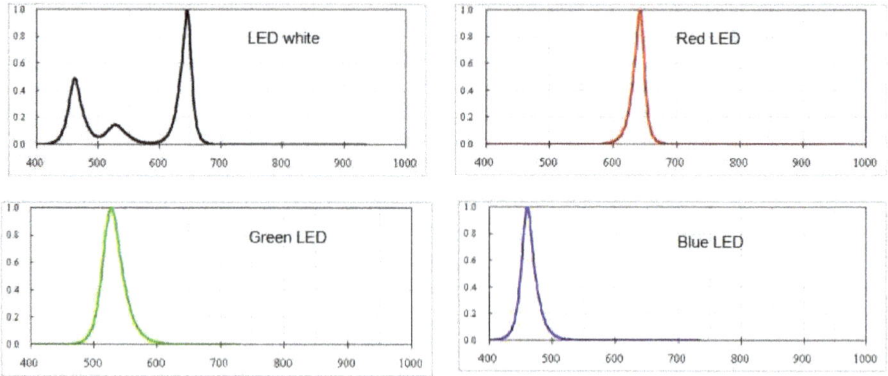

Figure 4. Spectrum distribution of LED light source.

3. Results and Discussion

The spectra of the light sources, including those being filtered by the individual colour filter, were measured by a commercial off-the-shelf spectrometer (FDSP with spectral range 380–1050 nm, ETA-Optik). It is clear in Figure 2 that the spectrum of the white light of the metal-halide lamp has three sharp peaks with wavelengths at 440 nm, 570 nm, and 590 nm respectively, in addition to being wide band. Hence, after colour filtering, the wavelengths of the metal-halide lamp are 680 nm, 580 nm, 555 nm, 550 nm, and 445 nm for red, orange, yellow, green, and blue filters respectively. On the other hand, the quartz-halogen lamp of Figure 3, the spectrum is more widely distributed than that of metal-halide lamp of Figure 2. Whereas, the peaks after filtering by the red, orange, yellow, green, and blue filters occur at 690 nm, 640 nm, 555 nm (above), 560 nm, and 450 nm respectively. As these filters were designed following the Gaussian transmission curve [13,14] to emulate the narrow spectral distribution of monochromatic LEDs, the filtered spectrum distributions conform to the filter specifications listed in Table 1. That is, they have a relatively focused wavelength to improve the image quality. However, for the quartz-halogen lamp, a considerable amount of noise still exists for those filtered by the red, yellow, and blue filters, especially the blue colour. In contrast, the red, green, and blue LEDs as Figure 4, the peaks occur at wavelength of 660 nm, 530 nm, and 460 nm respectively, which are close to the maximum spectral response of the TDI CCD in the range of 460 nm to 580 nm. The result demonstrates the sharpness of the focused wavelength of the monochrome LED without noise.

Typical optical images of the third and fourth layers are shown in Figures 5 and 6, respectively. Figure 5 illustrates the optical images obtained by using various light sources for the electrodes of the third layer source/drain (S/D) regions. The red light of both the filtered metal-halide lamp and the filtered quartz-halogen lamp results in blurred images. This is due to the spectrum is far from the maximum spectral response of the TDI CCD which is in the range of 460 nm to 580 nm. In contrast, the performance of the green light of both the filtered metal-halide lamp and the filtered quartz-halogen lamp is almost same as that of the green LED due to the similar spectrum among them. However, none of the green light gives the detailed metal traces exhibited by the blue light. For the blue light, the performance of the filtered metal-halide lamp is similar to that of the LED because of their narrow spectrum and the closeness to the maximum sensitivity of the TDI CCD. Whereas, the image quality of the blue light of the filtered quartz-halogen lamp is very poor due to its non-uniform and jagged spectrum. Hence, in terms of image quality of the third layer, the performance of the filtered metal-halide lamp is similar to that of the corresponding LED. However, the metal-halide lamp is relatively expensive, is very sensitive to the applied voltage, and its lighting intensity varies from lamp to lamp, in addition to its non-environmentally friendly characteristic. Thus, the blue LED is the most suitable light source for the third layer electrodes (marked by the tick symbol in Figure 5).

Figure 6 displays the optical images for the fourth layer TFT electrodes obtained by using various light sources. Overall comparison of the results show that the performance of the metal-halide lamp, the quartz-halogen lamp, and the LED is approximately the same, except that of the blue LED. The blue LED outperforms all of the other light sources. That is, the blue LED gives the clearest image (marked by a tick symbol) due to its short wavelength and better spatial resolution.

From the spectrum distribution shown in Figure 2, it is clear that the blue light of the filtered metal-halide lamp has a relatively clear peak at around 440 nm. However, it also contains wavelengths from 440 nm to 500 nm with a relatively constant proportion of about 25%. Hence, its performance is close to that of the blue LED for the third layer electrodes. But its performance for the fourth layer deteriorates due to scattering of the SiNx passivation.

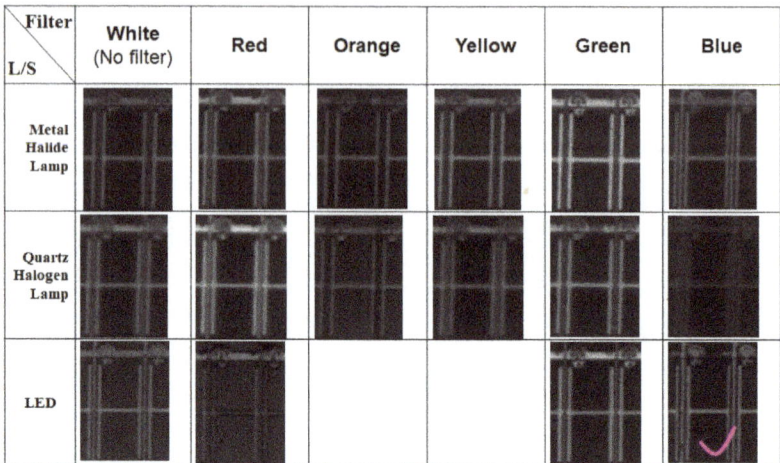

Figure 5. The third layer circuit pixel by various lamps for optical detection in TFT.

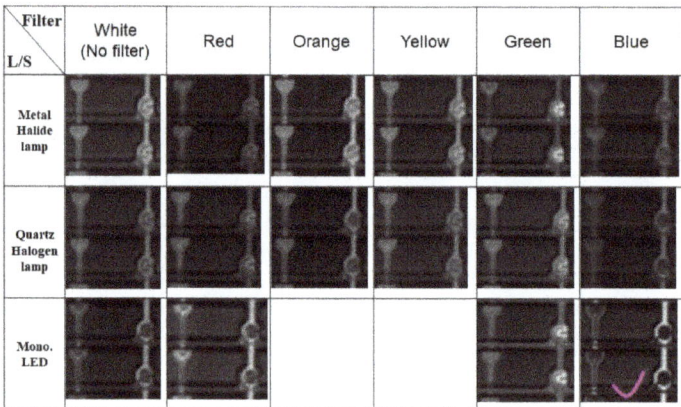

Figure 6. The fourth layer circuit pixel by various lamps for optical detection in TFT.

The images shown in Figures 5 and 6 indicate that the blue LED gives the best images for the third and fourth layer electrodes. Moreover, the image quality depends on both monochromatic and wavelength of the light source, not just either of them. Thus, in both Figures 5 and 6, no images are presented for the orange and yellow LEDs because they are not readily available. And more importantly they are not monochromatic.

Wagatsuma [15] mentioned that the emitted optical spectrum of chromium illuminated by an argon glow discharge plasma is in the wavelength range of 200–440 nm [15]. Thus, the blue LED responds well to Cr to trigger the TDI CCD, and results in a better image than other light sources. Recently, the panel pixels with low-temperature polysilicon used for in-plane switching LCDs and organic LEDs have been pushed to very small dimensions. Therefore, a shorter wavelength for AOI detection is essential to the shrinking electrode dimensions of high definition LCD panels which require better spatial resolution for detection. In contrast, the image by the red LED light is the poorest among the three LED light colours because its wavelength is farther from the range of 460 nm to 580 nm of the HS 8K TDI CCD. In other words, compatibility between the sensor and the light source is critical to the success of an AOI system, in addition to spatial resolution. As a further illustration of the best performance of the blue LED, Figure 7 shows the images for the source and drain of the third layer and the contact passivation of the fourth layer when illuminated by LEDs. The performance in descending order are blue LED> green LED > red LED, consistent with the above results. In other words, the blue LED gives the clearest distinct images of sharp boundaries. Hence, by combining the results of the third and fourth layer, the blue LED has the best performance.

Moreover, the present monochromic LEDs consume 90 Wp, which is about 36% of 250 Wp of both quartz-halogen and metal-halide lamps. Furthermore, the LEDs typically have a lifespan of 50,000 h, much longer than 2000 h of quartz-halogen lamps and 6000 h of metal-halide lamps. In other words, the cost-per-performance of the blue LED is superior to other light sources examined in this study.

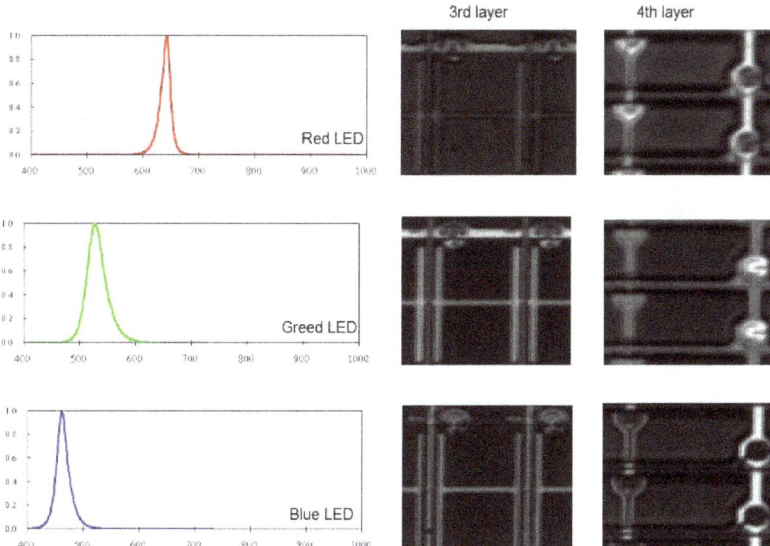

Figure 7. Comparison of third and fourth layer electrode images in TFT by LEDs.

4. Conclusions

This study experimentally investigates the effectiveness of various light sources on scanning the electrode pixels of TFTs using TDI CCD. The results show that blue LEDs provide the clearest images of both the third and fourth layer electrodes. Hence, it is the most suitable light source because of its spectrum compatibility with the TDI CCD system and better spatial resolution due to it short wavelength. The cost-per-performance provided by the blue LED is superior compared to other light sources typically used in such studies. Since modern monochromic LEDs consume 90 W_p, which is ~36% of the 250 W_p required for both quartz-halogen lamp and metal-halide lamp, LEDs are clearly more energy efficient. Together with the relatively long life span and ecological friendliness, LEDs are viable light sources for AOI. This is especially true for the blue LED for its spectrum compatibility with the TDI CCD sensor and its better spatial resolution. In other words, it should be adopted in AOI for both energy and performance considerations.

Author Contributions: Data curation, F.M.T.; Formal analysis, F.M.T.; Methodology, F.M.T.; Validation, J.H.C.; Writing—original draft, F.M.T.; Writing—review & editing, F.M.T. and J.H.C.

Funding: This research received no external funding

Conflicts of Interest: The authors declare no conflict of interest

References

1. Liu, S.; Wang, D.; Yang, Z.K.; Feng, X.; Sun, X.; Qiu, Y.; Dong, X. Key technology trends analysis of TFT-LCD. *Chin. J. Liq. Cryst. Disp.* **2018**, *33*, 457–463.
2. Li, X.H.; Bao, J.P.; Xu, B.; Fan, H.Y. Improvement research of TFT-LCD module black uniformity. *Chin. J. Liq. Cryst. Disp.* **2018**, *33*, 271–276.
3. Tzu, F.M.; Chou, J.H. Non-uniformity evaluation of flat panel display by automatic optical detection. In Proceedings of the 2016 11th International Microsystems, Packaging, Assembly and Circuits Technology Conference (IMPACT), Taipei, Taiwan, 26–28 October 2016; pp. 168–171.
4. Hu, S.; Fang, Z.; Ning, H.; Tao, R.; Liu, X.; Zeng, Y.; Yao, R.; Huang, F.; Li, Z.; Xu, M.; et al. Effect of Post Treatment For Cu-Cr Source/Drain Electrodes on a-IGZO TFTs. *Materials* **2016**, *9*, 623. [CrossRef] [PubMed]
5. Lee, C.H.; Sazonov, A.; Nathan, A. High hole and electron mobilities in nanocrystalline silicon thin-film transistors. *J. Non Cryst. Solids* **2006**, *352*, 1732–1736. [CrossRef]
6. Powell, M.J.; Glasse, C.; Curran, J.E.; Hughes, J.R.; French, I.D.; Martin, B.F. A fully self-aligned amorphous silicon TFT technology for large area image sensors and active-matrix displays. In *A Fully Self-Aligned Amorphous Silicon TFT Technology for Large Area Image Sensors and Active-Matrix Displays*; Schropp, R., Branz, H.M., Hack, M., Shimizu, I., Wagner, S., Eds.; Amorphous and Microcrystalline Silicon Technology-1998; Cambridge University Press: Cambridge, UK, 2011.
7. Grondzik, W.T.; Kwok, A.G.; Stein, B.; Reynolds, J.S. *Mechanical and Electrical Equipment for Buildings*; John Wiley & Sons Press: Hoboken, NJ, USA, 2014.
8. Kumar, T.S.; Harikumar, G.; Halpeth, M.K. *Light Right: A Practising Engineer's Manual on Energy-Efficient Lighting*; TERI Press: New Delhi, India, 2004; pp. 19–20.
9. Groß, R.; Bunke, D.; Gensch, C.O.; Stéphanie, Z.; Manhart, A. *Study on Hazardous Substances in Electrical and Electronic Equipment, Not Regulated by the RoHS Directive*; Contract No. 070307/2007/476836/MAR/G4; Öko-Institut e.V.: Freiburg, Germany, 2008; pp. 1–273.
10. Chulkov, A.O.; Vavilov, V.P.; Malakhov, A.S. A LED-based thermal detector of hidden corrosion flaws. *Russ. J. Nondestruct. Test.* **2016**, *52*, 588–593. [CrossRef]
11. Singh, P.; Sakarvadiya, V.; Dubey, N.; Kirkire, S.; Thapa, N.; Banerjee, A. Electrical coupling in multi-array charge coupled devices. In Proceedings of the SPIE Asia-Pacific Remote Sensing, Earth Observing Missions and Sensors: Development, Implementation, and Characterization Iv, New Delhi, India, 2 May 2016.
12. Luo, Y.C.; Smith, C.; Nixon, O.; Ledgerwood, M.; Kullar, S. High performance multispectral TDI CCD image sensors. In Proceedings of the SPIE Remote Sensing, sensors, systems, and next-generation satellites Xvii, Dresden, Germany, 24 October 2013.
13. Janecki, D. Gaussian filters with profile extrapolation. *Precision Eng.* **2011**, *35*, 602–606. [CrossRef]
14. Niconoff, G.M.; Torres-Rodriguez, M.A.; Morales, M.V.; Garcia, S.I.D.; Vara, P.M.; Carbajal-Dominguez, A. Generation of long-range curved-surface plasmonic modes and their propagation through thin metal films in a tandem array. *Appl. Opt.* **2017**, *56*, 8996–8999. [CrossRef] [PubMed]
15. Wagatsuma, K. Wavelength table of chromium emission lines in argon glow discharge optical emission spectrometry. *Fresenius. J. Anal. Chem.* **2000**, *367*, 414–415. [CrossRef]

© 2018 by the authors. Licensee MDPI, Basel, Switzerland. This article is an open access article distributed under the terms and conditions of the Creative Commons Attribution (CC BY) license (http://creativecommons.org/licenses/by/4.0/).

Article

Optical Detection of Green Emission for Non-Uniformity Film in Flat Panel Displays

Fu-Ming Tzu [1],* and Jung-Hua Chou [2]

1 Department of Marine Engineering, National Kaohsiung University of Science and Technology, Kaohsiung 80543, Taiwan
2 Department of Engineering Science, National Cheng Kung University, Tainan 70101, Taiwan; jungchou@mail.ncku.edu.tw
* Correspondence: fuming88@nkust.edu.tw; Tel.: +886-7-810-0888 (ext. 25245)

Received: 9 September 2018; Accepted: 5 November 2018; Published: 8 November 2018

Abstract: Among colours, the green colour has the most sensitivity in human vision so that green colour defects on displays can be effortlessly perceived by a photopic eye with the most intensity in the wavelength 555 nm of the spectrum. With the market moving forward to high resolution, displays can have resolutions of 10 million pixels. Therefore, the method of detecting the appearance of the panel using ultra-high resolutions in TFT-LCD is important. The machine vision associated with transmission chromaticity spectrometer that quantises the defects are explored, such as blackening and whitening. The result shows the significant phenomena to recognize the non-uniformity of film-related chromatic variation. In contrast, the quantitative assessment illustrates that the just noticeable difference (JND) of chromaticity CIE *xyY* at 0.001 is the measuring sensitivity for the chromatic variables (*x*, *y*), whereas JND is a perceptible threshold for a colour difference metric. Moreover, an optical device associated with a ^{198}Hg discharge lamp calibrates the spectrometer accuracy.

Keywords: optical; green; colour difference; chromaticity; just noticeable difference

---

1. Introduction

Currently, liquid crystal (LC) flat panel displays (FPDs) are moving toward high-imaging resolution, quick in-plane switches, vivid colour, saving energy, and low radiation [1,2]. For example, image resolution is advancing from high definition (HD) to ultra-high definition (UHD), i.e., from 2 K (1920 × 1080 pixels), to 4 K (3840 × 2160 pixels), 8 K (7680 × 4320 pixels), 16 K (15,360 × 8640 pixels), and even beyond to 32 K (30,720 × 17,280 pixels) [3]. Thus, full high-resolution images for enriching the stereoscopic visibility of the FPDs can be achieved [4,5]. To assure the image quality of the displays with such a high-resolution, non-destructive, automatic optical inspection (AOI) using photo sensors to detect defects is necessary so that quantitative assessment can be made instead of the subjective measurement by the human eyes. The practice of using human assessors to detect FPD defects is still popular in the liquid crystal display (LCD) industry. As the maximum light sensitivity of human eyes is the green light spectrum of around 555 nm [6], the present study focuses on this light spectrum to explore the possibility of using an automatic optical inspection (AOI) system to replace human assessors.

With the advances in both computer hardware and software, especially the image processing algorithms, image processing for defect detection in LCDs is getting popular in the research community and various approaches have been proposed [7]. Kuo et al. [8] employed image processing and neural networks to detect surface defects of colour filters to prevent losses arising from incorrect detection. Nam et al. [7] examined the defects in LCDs by utilizing the colour space LAB2000HL to replace human inspection to avoid person-to-person variations. Bin et al. [9] applied the level set method mura [10] defection which still relies heavily on the assessor's perception at the present time.

For AOI, images are typically captured by charge couple devices (CCDs). In general, two types of CCDs are commonly used in the industry: area and line scan. The fast area CCDs are more suitable for small areas to avoid image distortion; whereas the slower line scan CCDs are more suitable for large areas. As the panel size employed in this study is 1500 mm by 1850 mm (i.e., 6th generation), line-scan CCDs were selected to cope with the large area.

Among the line scan CCDs, the time-delay-integration (TDI) CCD was chosen because of its capability of multi-scan at one time and being able to accumulate the multiple exposures of moving objects effectively to enhance image quality. Moreover, the TDI CCD acquires the image with the pixels in synchronization continuously with the moving objects. Thus, all of the faint images of the same object becomes a high contrast and clear image in the end.

This study utilizes the just noticeable difference (JND) as the detection criterion according to that of the International Commission on Illumination (CIE).

2. Methodology

The TDI CCD adopted was a commercial off-the-shelf type, HS 8 K TDI CCD (Progressive, Piranha HS 8 K 68 kHz, TELEDYNE DALSA), used for its fast responsivity compared to other lines of CCDs. Its photo sensor offers the scanning mode under low light and slow speed during the TDI mode. The photo sensor grabs an image of a moving object while transferring the charge in synchronous scanning with the object (scanning image synchronization). The light source was an illuminant C with wavelengths including the ultra-ultraviolet (UV), visible, near infrared (NIR), and infrared (IR) range. Thus, this light can be reflected by various colours to be triggered and captured by the line CCD.

Currently, manual optical inspection (MOI) is widely used to observe the non-uniformity of colour filters by human eyes that identify diversified non-uniformity through various light sources, including fluorescent lights, halide lamps, sodium lamps, and light-emitted diodes (LED) as illustrated in Figure 1 (left), where CF denotes colour filter. The main drawback of MOI is its dependence on the human subjective judgement even though human eyes are very sensitive to colour changes. In contrast, the machine vision of AOI, shown in Figure 1 (right), is quantitative without humans' drawback.

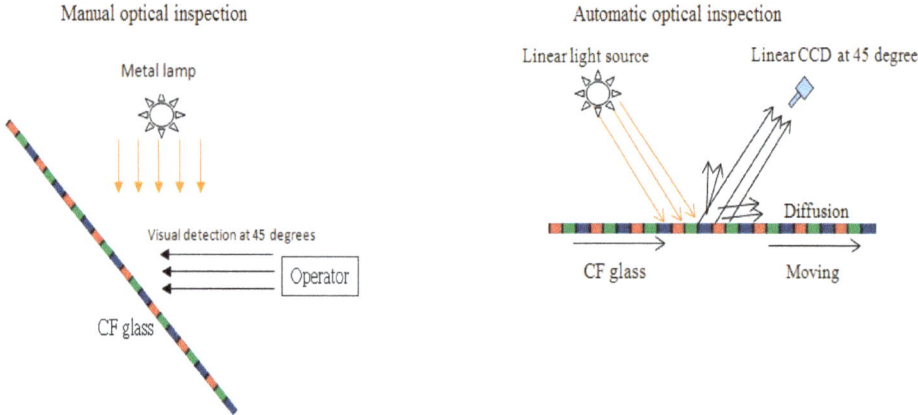

Figure 1. Illustration of optical inspection, manual optical inspection (MOI) (**left**) vs automatic optical inspection (AOI) (**right**) [3,10].

The characteristics of non-uniformity due to chromaticity or thickness difference can be typically inspected by the related grey level variant. An edge detection method can be applied to compare the grey level between the background and the selected area. Then, the features of the binary image of the segmented region is compared with those in the database. Thus, the defects can be determined through proper thresholds.

In the progress of colour image extraction, several approaches and models have been developed for colour judgement and applied for colour differences. Among these, in relation to FPDs, the tristimulus method, established by the International Commission on Illumination (CIE), is very popular and has been applied to a diverse colour space through non-contact optical measurements, including colour gamut, colour shift, and chromaticity difference. The formulas for the CIE XYZ colour space are as follows:

$$X = F \int_{380}^{780} T(\lambda)\, S(\lambda)\, \bar{x}(\lambda)\, d\lambda \tag{1}$$

$$Y = F \int_{380}^{780} T(\lambda)\, S(\lambda)\, \bar{y}(\lambda)\, d\lambda \tag{2}$$

$$Z = F \int_{380}^{780} T(\lambda)\, S(\lambda)\, \bar{z}(\lambda)\, d\lambda \tag{3}$$

$$F = \frac{100}{\int_{380}^{780} T(\lambda)\, S(\lambda)\, \bar{y}(\lambda)\, d\lambda} \tag{4}$$

In the above equations, CIE XYZ presents the tristimulus colour value which can be obtained through the spectrometer measurements. $T(\lambda)$ indicates the transmission spectrum and $S(\lambda)$ is a radiation profile for the illuminant C. Among the various colour systems, the CIE standard takes the spectrum response from the tristimulus values X, Y, and Z with the spectral matching functions $x(\lambda)$, $y(\lambda)$, and $z(\lambda)$ to obtain the normalised chromaticity coordinates x, y, and z. By tristimulus values X, Y, Z, the chromaticity coordinates x, y, and z are obtained as follows [11]:

$$x = \frac{X}{X + Y + Z} \tag{5}$$

$$y = \frac{Y}{X + Y + Z} \tag{6}$$

$$z = \frac{Z}{X + Y + Z} \tag{7}$$

The colour difference ΔE is designed to distinguish the perceived colours quantitatively to judge colour deviation [12,13] and is generally used to classify various visibility levels to reflect the perceivable degree of colour difference by certain criteria [14]. ΔE is typically expressed in terms of the Euclidean distance and is an index of visual perceptibility between the background and foreground. Its threshold is determined through repeated measurements. It is treated as the perceptual analogy of colour appearance for human vision.

Furthermore, CIE presents the colour distance by the metric ΔE^*ab, which occasionally is referred to as ΔE^*, dE^*, dE, or "Delta E". The perceptual non-uniformities in the CIELAB colour space have led CIE to refine the definition over the years, leading to CIE1994 and CIEDE2000. These non-uniformities are important because human eyes are more sensitive to certain colour than others. A good metric should take this into account in order for the notion of "just noticeable difference" (JND) to be meaningful. Otherwise, a certain ΔE may be insignificant in one part of the colour space while being significant in some other part. However, currently, the criterion of the JND value for the colour difference to be just noticeable is not set universally, although in practice, the JND value of 1.0 is often used. Mahy et al. [15] studied and evaluated a JND value of 2.3 ΔE in 1994. On the other hand, in the CIELAB colour space, the non-uniformity of perception is taken into account to reduce the inconsistency.

Berns [16] proposed the most prevalent methods to classify ΔE_{ab}, according to the perceptibility and acceptability. Initially, the perceptibility threshold determined the magnitude of colour difference of JND; a JND value of less than 1 implied the imperceptibility for viewing side by side [17]. Afterward, the acceptability was classified by three levels of colour difference for three visibilities of imperceptible, hardly perceptible, and easy perceptible sections. Furthermore, Perez et al. [18] determined the 50:50% perceptibility threshold (PT) and 50:50% acceptability threshold (AT) for computer-simulated samples of human gingiva using CIEDE2000 and CIELAB colour difference formulas. As a result, the PT and AT for CIEDE2000 and 95% confidence intervals were 1.1 and 2.8, respectively; the corresponding CIELAB values were 1.7 and 3.7. Nussbaum [17] proposed that two colour samples could be classified using ΔE_{ab} of less than 0.2 as "non-visible", between 0.2 and 1.0 as "very small visual", between 1.0 and 3.0 as "small", between 3.0 and 6.0 as "medium", and greater than 6.0 as "large".

The JND is a quantitative index to describe the minimal amount of variation in a stimulus perceived by an observer; it has a statistical nature. In the display industry, the CIE xyY standard colour systems usually adopt the spectrum response from the tristimulus values X, Y, and Z that are used to obtain the normalised chromaticity coordinates x, y, and z. Through the transformation of the CIE xyY colour coordinates, the coordinates (a, b) in the CIELAB uniform colour space are as follows:

$$L = 116 \left(\frac{Y}{Y_n}\right)^{1/3} - 16 \tag{8}$$

$$a = 500 \left[\left(\frac{X}{X_n}\right)^{1/3} - \left(\frac{Y}{Y_n}\right)^{1/3}\right] \tag{9}$$

$$b = 200 \left[\left(\frac{Y}{Y_n}\right)^{1/3} - \left(\frac{Z}{Z_n}\right)^{1/3}\right] \tag{10}$$

The symbols X_n, Y_n, and Z_n express the constant for the daylight source. The equations for colour difference are as follows:

$$\Delta E_{1976} = \sqrt{(\Delta L)^2 + (\Delta a)^2 + (\Delta b)^2} \tag{11}$$

$$\Delta E_{1994} = \sqrt{\left[\frac{\Delta L}{K_L S_L}\right]^2 + \left[\frac{\Delta C}{K_C S_C}\right]^2 + \left[\frac{\Delta H}{K_H S_H}\right]^2} \tag{12}$$

$$\Delta E_{2000} = \sqrt{\left(\frac{\Delta L'}{K_L S_L}\right)^2 + \left(\frac{\Delta C'}{K_C S_C}\right)^2 + \left(\frac{\Delta H'}{K_H S_H}\right)^2 + R_T \left[\frac{\Delta C'}{K_C S_C}\right]\left[\frac{\Delta H'}{K_H S_H}\right]} \tag{13}$$

In the above equations, ΔL, Δa, and Δb are the difference between the test and reference specimens in lightness, redness or greenness, and yellowness or blueness, respectively. The weighting factor K depends on the specific application; S_L, S_C, and S_H are the compensation factors for lightness, chroma, and hue, respectively; whereas $\Delta L'$, $\Delta C'$, and $\Delta H'$ are the specific lightness, chroma, and hue in ΔE_{2000}. It has been found that the colour space of the colour difference formula of CIELAB is not completely uniform.

Figure 2 presents a schematic of the proposed architecture that detects the spot defects onto the LCD panel and measures the chromaticity of the panel. The conveyor reduces the tact time to increase the throughput of the production line. The system will send out an alarm message when a defect is detected.

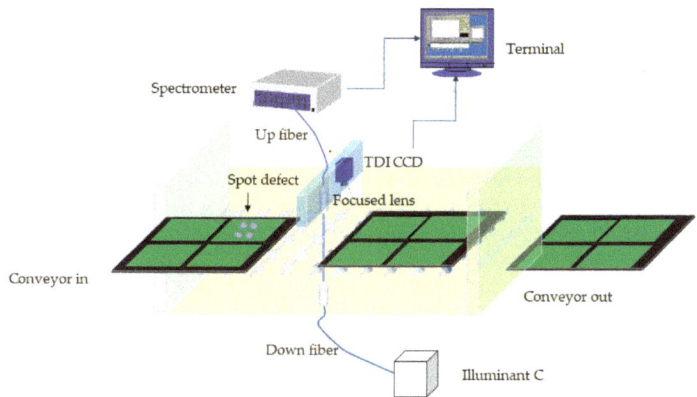

Figure 2. An installation for AOI engaged chromatic measurement by optical device.

3. Experimental Procedure

The present AOI method utilizes a VIS transmission chromatic spectrometer, Etaoptik (with a wavelength of 380 nm~1050 nm, focal length of 50 mm, detector: Si diode line array) to detect the non-uniformity defects on the green emission layer of the colour filter of the 6th generation (1500 mm × 1850 mm) TFT-LCD panels. The experiments were conducted in the class 1000 clean-room at 25 °C. The symmetric architecture uses a photo sensor linking image-grab card to acquire the two-dimensional image information. The line-scan TDI-CCD composes the multi array of the pixel sensors, DALSA HS-80-08K40 (dynamic range of 56 dB, line rates up to 34 kHz and throughput up to 320 MHz). The large panel substrate moved with constant speed for the whole sample scanning. The illuminant C source illuminated the area covered by the TDI-CCD.

Typically, four kinds of spot non-uniformity defects on the green layer were to be evaluated. They are labelled as A_1, A_2, B_1, and B_2 in Figure 3. Both A_1 and A_2 were the samples with artificial defects at 20 mm in diameter, which were dark regions and thick films of low transparency. In contrast, both B_1 and B_2 (also manmade defects of the same diameters of 20 mm) were bright regions and thin films with high transparency. After obtaining the image data, the embedded software identified the grey level variations to judge whether the threshold was reached to reveal the defect.

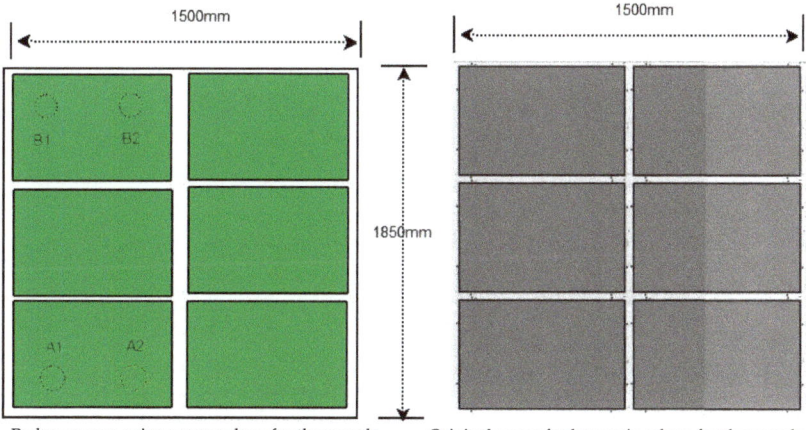

Figure 3. Defect locations of the 6th generation in TFT-LCD panel.

4. Results and Discussion

Figure 4 shows the different spectral distributions for emissive layers of red, green, blue, and black matrices (BMs) through the spectrometer. The spectrum analysis is by 0.8 nm interval wavelength (BTC611E, back-thinned CCD array, working wavelengths from 300 nm to 1050 nm, produced by B&W TEK). Among these photo resists (PRs), the optical response of green layer reaches 65%, i.e., having larger power intensity than other PRs.

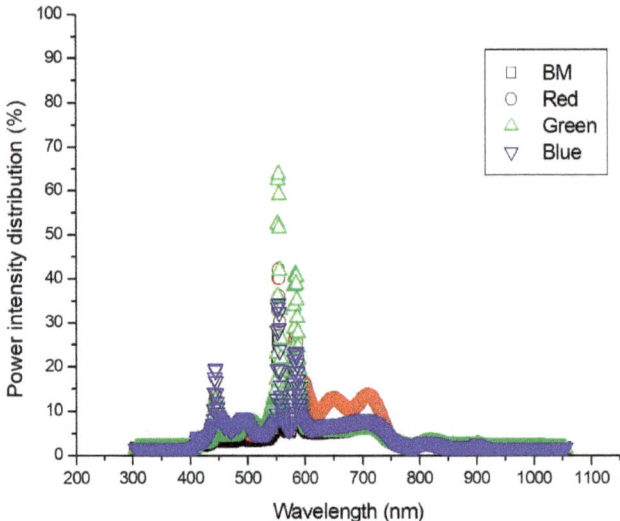

Figure 4. Spectral profile for red, green, blue, and black matrices, respectively [3].

Figure 5 depicts the CIE x profiles across the spot defect at intervals of 1 mm for A_1 and A_2. This task identifies the chromatic tendency versus transparency. As shown, both A_1 and A_2 have a concave variation due to the dark region as well as the low transparency. Thus, CIE x on green emission layer exhibits the concave variation in the dark region.

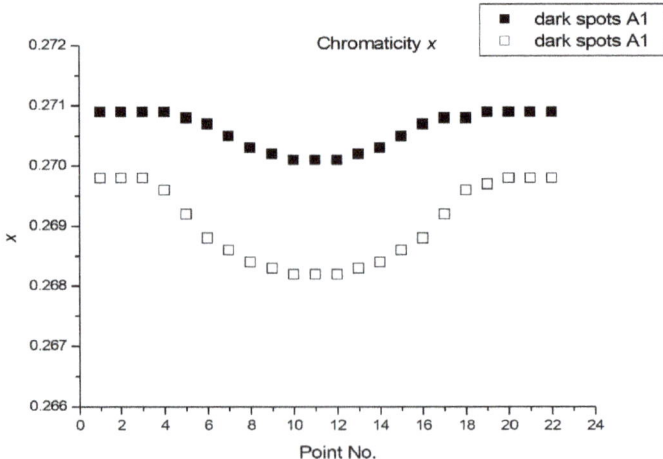

Figure 5. Variation of CIE x for dark spots defect of A_1 and A_2.

Figure 6 shows the CIE x profiles across the spot defect at intervals of 1 mm for B_1 and B_2. As shown, both B_1 and B_2 have a convex variation as they are bright region and have high transparency. Thus, CIE x on green emission layer shows the convex variation in the bright region, and vice versa.

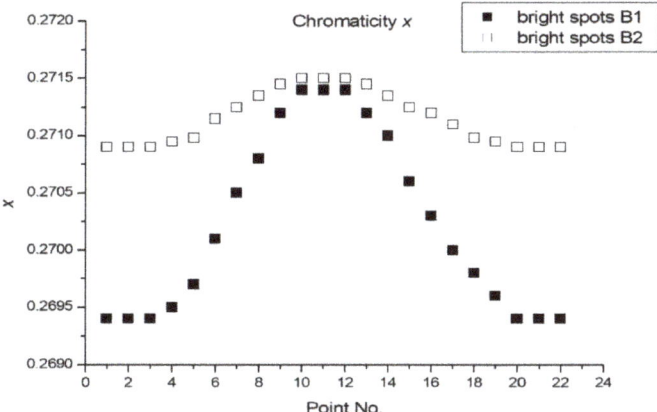

Figure 6. Variation of CIE x for bright spots of B_1 and B_2.

Figure 7 illustrates that the CIE y variation for samples A_1 and A_2 (dark region and low transparency) have a convex characteristic. The CIE y demonstrates a proportional film thickness; a larger chromaticity with a thicker film and vice versa [14]. For the evaluation of colour variation, the chromaticity of CIE y is also used to judge the blue layer film in the flat panel industry [14,19].

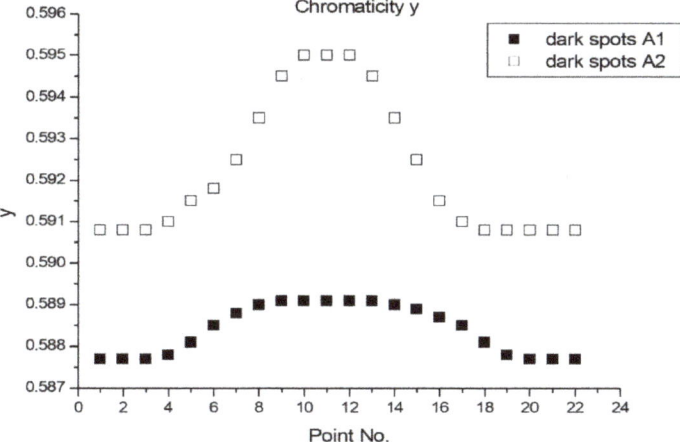

Figure 7. Variation of CIE y for dark spots of A_1 and A_2.

Figure 8 shows that the CIE y variation of samples B_1 and B_2 have a concave feature. The CIE y demonstrates a proportional film thickness; the larger chromaticity with the thicker film and vice versa. Hence, Figures 5–8 indicate the variation of colour saturation with film transparency. For dark defects A_1 and A_2, delta_x > 0 and delta_y > 0, the saturation increases; for bright defect B, the reserve is observed. This behavior is expected for a color film thickness variation since decreasing the film thickness not only increasing its transmission but also widening its spectral width; hence decreasing the color saturation [14]. In the limiting case, when the film thickness drops to zero, its transmittance becomes 1 and the saturation falls to 0, corresponding to a colorless full transparent region.

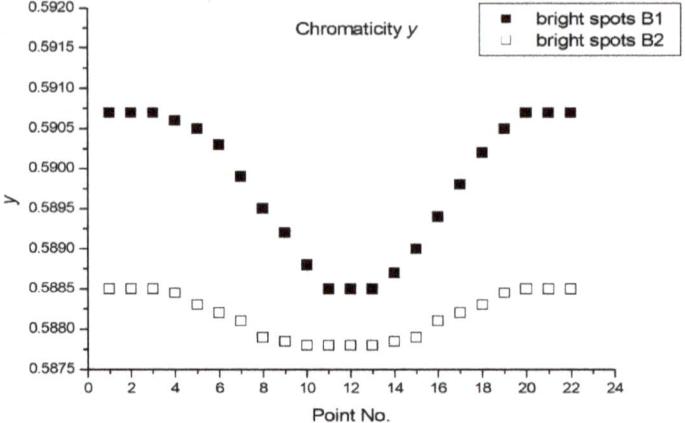

Figure 8. Variation of CIE y for bright spots of B_1 and B_2.

Figure 8. Variation of CIE y for bright spots of B_1 and B_2.

The CIE Y as shown in Figure 9 exhibits the luminance of the dark regions when the transmitted light passes through the region of spot non-uniformity. The transmitted light always decreases with thickness, regardless of the transparent material. The thin film has high transparency and the thick film has low transparency. It is obvious that the luminance of the transmitted light Y increases with transparency, being higher for thinner film regions (B defects as Figure 10) and lower for thicker regions (A defects as Figure 9). Thus both A_1 and A_2 have a concave variation with thick film, and both B_1 and B_2 are convex variation with thin film.

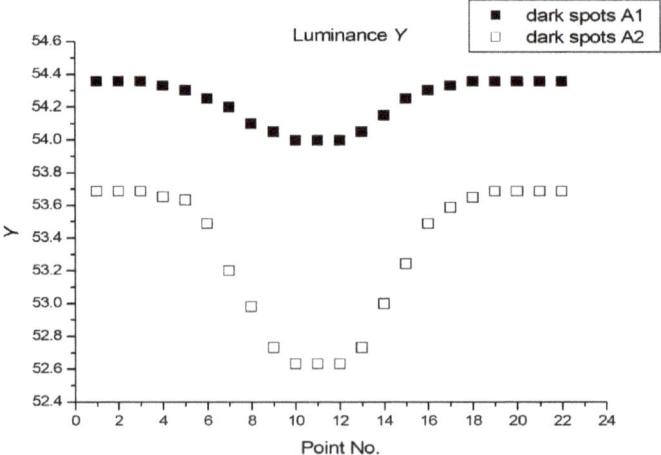

Figure 9. Variation of CIE Y for dark spots defects of A_1 and A_2.

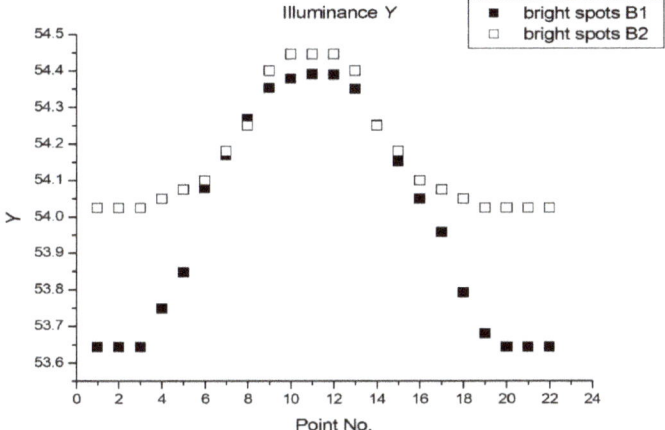

Figure 10. Variation of CIE *Y* for bright spots defect of B_1 and B_2.

Figure 11 presents the dark spots A_1 and A_2 of the original map and the enhanced grey map with the associated CIE *y* variation. The original map was acquired by the line-scan TDI-CCD in grey levels from 0 to 255. The enhanced map was obtained from the original map by rescaling the grey levels from 0–255 to 120–170 by Photoshop to increase the image contrast of the dark spots. Thus, the defects have sharper image than original map. Moreover, the maximum CIE *y* differences are 0.001 and 0.004 for spot A_1 and A_2, respectively. The profile indicates that CIE *y* increases with thick film [14].

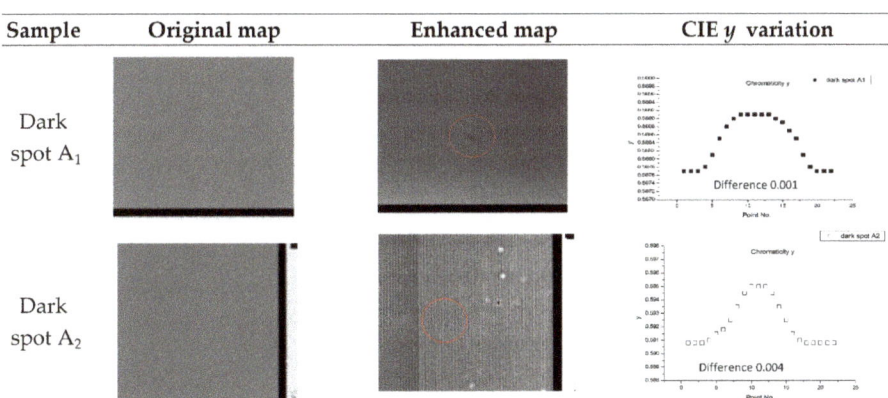

Figure 11. The original map and enhanced map illustrate CIE *y* variation for A_1 and A_2.

Figure 12 illustrates the original map and enhanced grey with respect to CIE y variations for the bright spots B_1 and B_2. The maximum differences in CIE *y* are 0.002 and 0.001 for B_1 and B_2, respectively. The same grey rescaling scheme as did for A_1 and A_2 was also performed here to obtain sharper images of the defects. The profile indicates that CIE *y* decreases as the film thickness decreases, similar to the observations of [14].

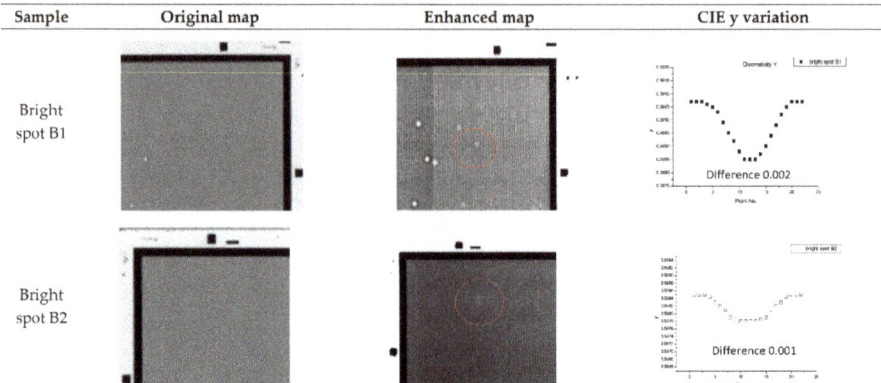

Figure 12. The original map and enhanced map illustrating defects by CIE y variations for B_1 and B_2, respectively.

Figure 13 presents the distribution of the colour gamut of the defects, A_1, A_2, B_1, and B_2 based on CIE xyY colour space. The triangle area uses the mathematical vector to depict the colour space. It shows that the imaging quality of colour saturation is very poor due the defects. Assuming the blue and red colours in the chromaticity are equivalent to the NTSC standard. Usually, the colour gamut can reveal the colour saturation, vivid, sharp, as well as contrast in the full colour of the display. The chart shows that these defects result in deficient colour saturation. As a result, the defects are more perceptual in the display.

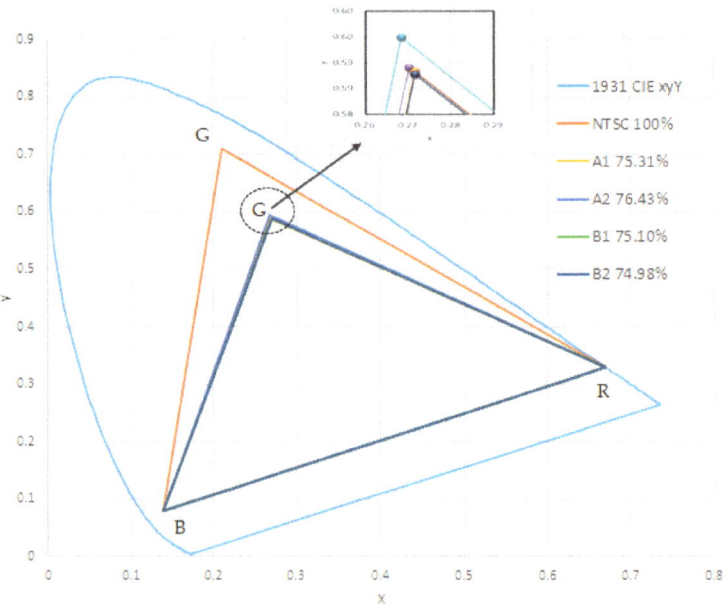

Figure 13. Colour gamut for A_1, A_2, B_1, and B_2.

Figure 14 presents the architecture of the calibrated spectrometer linking [198]Hg discharged lamp which utilizes spectral lines to calibrate the spectrometer in the visual spectrum [20–22]. This calibrated apparatus adopts the Model 6034 pencil lamp from Oriel Instruments. The criterion for the shift of the wavelength is small than 1 pixel for the spectrometer; 1 pixel calculation based on the spectrometer capacity is the difference between the maximum value and the minimum value divided by the spectrometer resolution. In this task, 1 pixel is at $(1050 - 380)/256 = 2.6$ nm. The measurement standard deviation is at 0.8 nm for the spectrometer.

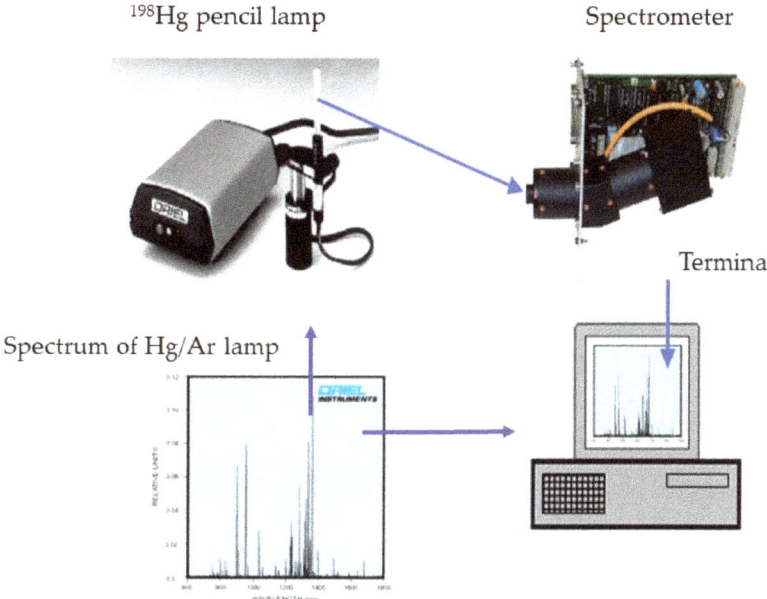

Figure 14. Measurement procedure of ^{198}Hg pencil lamp.

Table 1 lists the ranges of various CIE xyY for the examined defects. The results were obtained by repeating measurements ten times for each sample. The ranges of CIE x are 0.001, 0.002, 0.002, and 0.001 for samples A_1, A_2, B_1, and B_2, respectively. The corresponding ones for CIE y are 0.001, 0.004, 0.002, and 0.001; 0.4, 1.1, 0.7, and 0.4 for CIE Y. The maximum variation of CIE Y occurs in A_2.

Table 1. Statistics for colour variation for CIE xyY.

Samples	CIE x	CIE y	CIE Y
A_1	0.001	0.001	0.4
A_2	0.002	0.004	1.1
B_1	0.002	0.002	0.7
B_2	0.001	0.001	0.4

Table 2 tabulates the JND statistic according to colour difference referring to the formula of CIE1976, CIE1994, and CIEDE2000, indicating the visual intensity. Except for A_2, these defect samples are imperceptible by human eyes. The ΔE colour differences of A_2 are 3.3, 1.3, and 1.0 for CIE1976, CIE1994, and CIE2000, respectively, i.e., it is an obvious defect. The optical detected results show that the spot non-uniformity defects can be identified, even if the chromatic differences in CIE x and CIE y are below 0.001. In contrast, the colour gamut comparing to the 100% standard of NTSC shows that those of defects A_1, A_2, B_1, and B_2 are 75.31%, 76.43%, 75.10%, and 74.98%, respectively. That is, they result in very poor colour saturation; also, a high colour difference leads to a low colour gamut. With the sensitivity of the present AOI, it is more reliable than MOI as only A_2 is perceptible, whereas the others are hardly perceptible by human assessors. For completeness, Table 3 tabulates the ^{198}Hg discharge lamp with spectral lines to identify the standard deviation of spectral measurement to show the accuracy of the present AOI method.

Table 2. Colour differences based on 1976, 1994, and 2000 formula, and colour gamut.

Samples	$\Delta E1976_{JND2.3}$	$\Delta E1994_{JND1.0}$	$\Delta E2000_{JND1.0}$	Colour Gamut
A_1	0.9	0.4	0.3	75.31%
A_2	3.3	1.3	1.0	76.43%
B_1	1.8	0.7	0.5	75.10%
B_2	1.1	0.4	0.3	74.98%

Table 3. Standard deviation of the spectrometer by calibrated ^{198}Hg discharged lamp.

Measured (nm)	Lamp (nm)	Difference (nm)
404.077	404.656	−0.579
434.232	435.835	−1.603
545.529	546.074	−0.545
577.305	576.959	0.346
Standard deviation		0.797

5. Conclusions

An automatic optical method using a line-scan mode TDI-CCD with a transmitted spectrometer to detect non-uniformity of the green emission layer for ultra-high resolution TFT-LCDs is presented. The evidence shows a significant breakthrough to identify spot non-uniformity related to chromatic variation, and even JND is lower than the perceptibility threshold of a human assessor. Moreover, the CIE x of the green emission layer shows the concave variation for the dark region and the convex variation for the bright region. The CIE y is proportional film thickness, i.e., the larger chromaticity with the thicker film and vice versa. On the other hand, for CIE Y, both A_1 and A_2 have a concave variation with thick film, and both B_1 and B_2 have a convex variation with thin film. The present AOI has capacity of reaching the intensity of colour difference of 0.3 based on ΔE_{2000}, which is more sensitive than that of the JND. Moreover, it can detect all of the defects samples investigated, but MOI can only detect defect A_2; other defects are hardly perceptible by MOI. That is, the present method can quantify the defects accurately, and thus can substitute MOI in the display industry.

Author Contributions: Data curation, F.-M.T.; Formal analysis, F.-M.T.; Methodology, F.-M.T.; Validation, J.-H.C.; Writing–original draft, F.-M.T.; Writing–review & editing, F.-M.T. and J.-H.C.

Funding: This research received no external funding.

Conflicts of Interest: The authors declare no conflicts of interest.

References

1. Liu, S.; Wang, D.; Yang, Z.K.; Feng, X.; Sun, X.; Qiu, Y. Key technology trends analysis of TFT-LCD. *Chin. J. Liq. Cryst. Disp.* **2018**, *33*, 457–463.
2. Li, X.H.; Bao, J.P.; Xu, B.; Fan, H.Y. Improvement research of TFT-LCD module black uniformity. *Chin. J. Liq. Cryst. Disp.* **2018**, *33*, 271–276.
3. Tzu, F.M.; Chou, J.H. Non-uniformity evaluation of flat panel display by automatic optical detection. In Proceedings of the 11th International Microsystems, Packaging, Assembly and Circuits Technology Conference (IMPACT), Taipei, Taiwan, 26–28 October 2016; pp. 168–171.
4. Kwon, K.J.; Kim, M.B.; Heo, C.; Kim, S.G.; Baek, J.S.; Kim, Y.H. Wide color gamut and high dynamic range displays using RGBW LCDs. *Displays* **2015**, *40*, 9–16. [CrossRef]

5. Kim, D.U.; Kim, J.S.; Choi, B.D. A Low-Power Data Driving Method With Enhanced Charge Sharing Technique for Large-Screen LCD TVs. *J. Disp. Technol.* **2015**, *11*, 346–352. [CrossRef]
6. Bergman, L.; McHale, J.L. *Handbook of Luminescent Semiconductor Materials*; CRC Press: Boca Raton, FL, USA, 2016; pp. 268–270.
7. Nam, G.; Lee, H.; Oh, S.; Kim, M.H. Measuring Color Defects in Flat Panel Displays Using HDR Imaging and Appearance Modeling. *IEEE Trans. Instrum. Meas.* **2016**, *65*, 297–304. [CrossRef]
8. Kuo, C.F.; Hsu CT, M.; Fang, C.H.; Chao, S.M.; Lin, Y.D. Automatic defect inspection system of colour filters using Taguchi-based neural network. *Int. J. Prod. Res.* **2013**, *51*, 1464–1476. [CrossRef]
9. Ali AS, O.; Asirvadam, V.S.; Malik, A.S.; Eltoukhy, M.M.; Aziz, A. Age-Invariant Face Recognition Using Triangle Geometric Features. *Int. J. Pattern Recognit. Artif. Intell.* **2015**, *29*, 1556006.
10. Tzu, F.M.; Chou, J.H. Spot Mura evaluation in TFT-LCDs using automatic optical inspection. In Proceedings of the 5th International Microsystems Packaging Assembly and Circuits Technology Conference, Taipei, Taiwan, 20–22 October 2010; Volumes 1–4.
11. Gomez-Polo, C.; Munoz, M.P.; Lunego MC, L.; Vicente, P.; Galindo, P.; Casado, A.M.M. Comparison of the CIELab and CIEDE2000 color difference formulas. *J. Prosthet. Dent.* **2016**, *115*, 65–70. [CrossRef] [PubMed]
12. Strocka, D. Color difference formulas and visual acceptability. *Appl. Opt.* **1971**, *10*, 1308–1313. [CrossRef] [PubMed]
13. Gomez-Polo, C.; Munoz, M.P.; Luengo MC, L.; Vicente, P.; Galindo, P.; Casado, A.M.M. Comparison of two color-difference formulas using the Bland-Altman approach based on natural tooth color space. *J. Prosthet. Dent.* **2016**, *115*, 482–488. [CrossRef] [PubMed]
14. Tzu, F.M.; Chou, J.H. Slit-Mura Detection through Non-contact Optical Measurements of In-Line Spectrometer for TFT-LCDs. *IEICE Trans. Electron.* **2009**, *92*, 364–369. [CrossRef]
15. Mahny, M.; Vaneycken, L.; Oosterlinck, A. Evaluation of uniform color spaces developed after adoption of CIELAB and CIELUV. *Color Res. Appl.* **1994**, *19*, 105–121.
16. Berns, R.S. *Billmeyer and Saltzman's Principles of Color Technology*; Wiley: New York, NY, USA, 2000.
17. Nussbaum, P. Colour Measurement and Print Quality Assessment in a Colour Managed Printing Workflow. Doctoral Dissertation, The Norwegian Color Research Laboratory, Faculty of Computer Science and Media Technology, Gjøvik University College, Gjøvik, Norway, 2011; pp. 264–284.
18. Perez, M.M.; Ghinea, R.; Herrera, L.J.; Carrillo, F.; Ionescu, A.M.; Paravina, R.D. Color difference thresholds for computer-simulated human Gingiva. *J. Esthet. Restor. Dent.* **2018**, *30*, E24–E30. [CrossRef] [PubMed]
19. Lee, J.Y.; Yoo, S.I. Automatic detection of region-mura defect in TFT-LCD. *IEICE Trans. Inf. Syst.* **2004**, *87*, 2371–2378.
20. Martinsen, P.; Jordan, B.; McGlone, A.; Gaastra, P.; Laurie, T. Accurate and precise wavelength calibration for wide bandwidth array spectrometers. *Appl. Spectrosc.* **2008**, *62*, 1008–1012. [CrossRef] [PubMed]
21. Veza, D.; Salit, M.L.; Sansonetti, C.J.; Travis, J.C. Wave numbers and Ar pressure-induced shifts of Hg-198 atomic lines measured by Fourier transform spectroscopy. *J. Phys. B At. Mol. Opt. Phys.* **2005**, *38*, 3739–3753. [CrossRef]
22. Sansonetti, C.J.; Salit, M.L.; Reader, J. Wavelengths of spectral lines in mercury pencil lamps. *Appl. Opt.* **1996**, *35*, 74–77. [CrossRef] [PubMed]

© 2018 by the authors. Licensee MDPI, Basel, Switzerland. This article is an open access article distributed under the terms and conditions of the Creative Commons Attribution (CC BY) license (http://creativecommons.org/licenses/by/4.0/).

Article

Surface Treatments on the Characteristics of Metal–Oxide Semiconductor Capacitors

Ray-Hua Horng [1,2,*], Ming-Chun Tseng [3] and Dong-Sing Wuu [3]

1. Institute of Electronics, National Chiao Tung University, Hsinchu 300, Taiwan
2. Center for Emergent Functional Matter Science, National Chiao Tung University, Hsinchu 300, Taiwan
3. Department of Materials Science and Engineering, National Chung Hsing University, Taichung 402, Taiwan; idt266@yahoo.com.tw (M.-C.T.); dsw@nchu.edu.tw (D.-S.W.)
* Correspondence: rhh@nctu.edu.tw Tel: +886-3-5712121 (ext. 54138)

Received: 29 November 2018; Accepted: 17 December 2018; Published: 20 December 2018

Abstract: The properties of metal-oxide semiconductor (MOS) capacitors with different chemical treatments have been examined in this study. A MOS capacitor consists of an Al_2O_3/n-GaN/AlN buffer/Si substrate. Four chemical treatments, containing organic solvents, oxygen plasma and BCl_3 plasma, dilute acidic and alkali solvents, and hydrofluoric acid, were used to reduce the metal ions, native oxides, and organic contaminants. The n-GaN surface was treated with these chemical treatments before Al_2O_3 was grown on the treated n-GaN surface to reduce the interface state trap density (D_{it}). The value of D_{it} was calculated using the capacitance–voltage curve at 1 MHz. The D_{it} of a u-GaN surface was modified using various solutions, which further influenced the contact properties of GaN.

Keywords: chemical treatment; capacitor; interface state trap density

---

1. Introduction

Surface cleaning treatments are the foundation of a semiconductor device fabrication process [1,2]. Surface cleaning significantly affects the epitaxial defects [2], metal contact resistance/stability [3], and overall device quality of GaN-based devices [4]. Evaluating surface cleanliness requires considering the electrical properties of the device, structure, and interface state trap of the surface. Moreover, a surface treatment is used to remove the native oxides, organic contaminants, metal ions, particulates, residual species, and weaknesses in atomic bonding.

Recently, AlGaN/GaN high electron mobility transistors (HEMTs) were demonstrated for use in power electronic devices. In an HEMT device, a high saturation current, low leakage current, and high transconductance are necessary. Therefore, a low-resistance ohmic contact and low interface state trap density (D_{it}) must be obtained for an HEMT device. Interface states may cause various operational stability and reliability drawbacks in GaN-based HEMTs such as threshold voltage instability [4] and current collapse phenomena [5]. A surface treatment not only improves the device performance but also enhances the ohmic contact characteristics of GaN with metals [3].

In addition, surface treatments have been proposed to improve the ohmic contact properties between a low work function metal contact and a GaN or AlGaN surface. N vacancies are created during surface treatment and act as shallow donors for electrons and increase the surface doping concentration to overcome the Schottky barrier height for carrier transport. Therefore, improving the contact properties and reducing contact degradation are crucial for an AlGaN/GaN HEMT. A poor surface quality causes surface defects and contamination of the interface states, which originate from dangling bonds. The surface chemical treatment is not the only way to improve AlGaN/GaN HEMT device performance. The phosphorus-based annealing processes ($POCl_3$, P_2O_5) also modified surface conditions before dielectrics material deposition to further modify the AlGaN/GaN HEMT [6].

The metal-oxide-semiconductor HEMTs(MOS-HEMTs) structure using different dielectrics material causes different device performance of MOS-HEMTs, such as different C–V characteristics, specific resistance (R_{on}), breakdown voltage, D_{it} value, saturation drain current of the devices [7]. The MOS-HEMTs structure using Al_2O_3 [8] will be effective to enhance the breakdown voltage generated from the gate leakage. The ALD is a surface-controlled layer-by-layer process for the deposition of thin films with atomic layer accuracy. The Al_2O_3 used for MOS-HEMTs not only improve the basic electronic properties but also show low leakage current and high breakdown voltage. In this study, four chemical pretreatments were used for MOS capacitors before atomic layer deposition (ALD) of Al_2O_3 to modify the surface quality. The characteristics of MOS capacitors and the ohmic contact characteristics of GaN with the four chemical treatments are discussed.

2. Experiments

Figure 1 presents a schematic of an MOS capacitor. The MOS capacitor was grown on silicon (111) substrate through metal-organic chemical vapor deposition. The MOS capacitor consisted of an AlN nucleation layer, 2 μm GaN buffer layer, and 1 μm n-GaN layer Figure 1. Dilute HCl was used to remove the native oxides, and an organic solution was then used to remove organic contaminants in an ultrasonic cleaner before the MOS capacitor fabrication process. Mesa isolation was achieved using inductively coupled plasma-reactive ion etching with $BCl_3/Cl_2/Ar$ plasma, and the sample was subjected to four chemical pretreatments before Al_2O_3 oxide layer deposition. The four chemical pretreatments are shown in Table 1.

Table 1. The four chemical pretreatment descriptions.

Treatment	Description of Surface Treatment
1	ACE → IPA → DI Water → O_2 Plasma
2	ACE → IPA → DI Water → O_2 Plasma → $HCl:H_2O$ for 1 min → DI Water → $HF:H_2O$ for 1 min
3	ACE → IPA → DI Water → O_2 Plasma → BCl_3 Plasma → $HCl:H_2O$ for 1 min → DI Water → $HF:H_2O$ for 1 min
4	ACE → IPA → DI Water → O_2 Plasma → $HF:H_2O$ for 1 min → DI Water → $NH_4OH:H_2O$ for 1 min → DI Water → $HF:H_2O$ for 1 min → DI Water → $HCl:H_2O$ for 1 min → DI Water → $HF:H_2O$ for 1 min

Subsequently, a 50-nm-thick Al_2O_3 layer as a gate oxide was deposited using ALD at 300 °C under 6 mbar. In the ALD process, water vapor and trimethylaluminum were respectively used as O and Al sources, which were alternate pulse forms, resulting in the formation of the Al_2O_3 layer. Ti/Al/Ti/Au (25 nm/125 nm/45 nm/55 nm) alloyed for an ohmic contact with n-GaN was then formed through rapid thermal annealing (RTA) at 870 °C for 30 s in N_2 ambient. Finally, Ni/Au metal was deposited on the Al_2O_3 by using an E-gun evaporating system. The contact characteristics of Ti/Al/Ti/Au deposited on the GaN were evaluated by circular transmission line model (CTLM). The CTLM structure's inner and outer circle were 50, 100, 150, 200, and 250 μm, respectively. The inner radius of the pad was 100 μm.

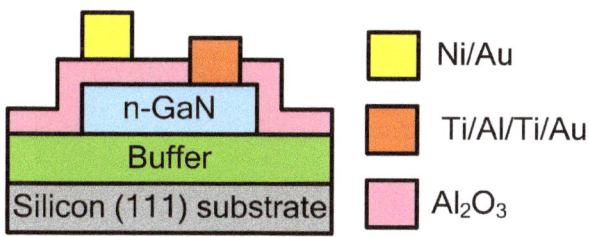

Figure 1. Schematic of an MOS capacitor.

The chemical bonding states on the GaN surface were characterized using X-ray photoelectron spectroscopy (XPS) with a monochromate Al $K\alpha$ X-ray (energy; 1486.6 eV). The shift in the XPS spectra was calibrated using a charge neutralization gun because of surface charge accumulation by emitting photoelectrons. The angle between the incident photons and the detected photoelectrons was set at 45°, which is sensitive to an analysis of surface chemical states.

3. Results and Discussion

The hysteresis behavior of capacitance–voltage (C–V) curves is strongly correlated to the trap density at the GaN/Al$_2$O$_3$ interface of a MOS capacitor. Figure 2 shows the hysteresis behavior of the C–V curves of MOS capacitors treated with various chemical treatments. Obviously, the MOS capacitor treated with treatment 1 exhibited the largest amount of hysteresis. Treatment 1 consisted of an organic solvent and O$_2$ plasma, which were used to remove particles from the air ambiance, and stripped residual photoresist and organic contaminants. The dangling bond, weaknesses in atomic bonding, and native oxides are difficult to remove using treatment 1. Generally, they are removed by complexes composed of organic and inorganic solvents, such as treatments 2–4.

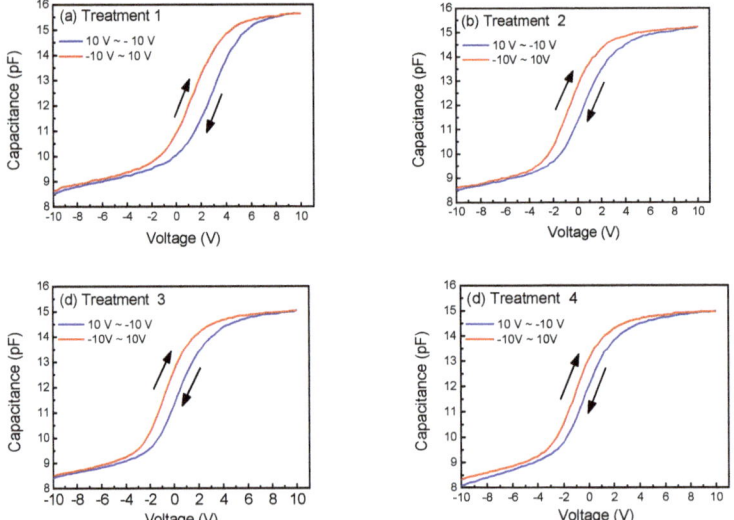

Figure 2. The Capacitance–voltage curves of MOS capacitors treated with (**a**) treatment 1, (**b**) treatment 2, (**c**) treatment 3, and (**d**) treatment 4.

To calculate the D_{it} value, the oxide capacitance (capacitance at accumulation, C_{ox}) was measured using the C–V curves. The flat band voltage (V_{FB}) was calculated using Equation (1) to quantify the relative shifts for analyzing the hysteresis behavior of the capacitors; V_{FB} was measured using the C–V curves at a point of C_{FB} obtained using [9]:

$$C_{FB} = \frac{C_{ox}\varepsilon_s A/\lambda}{C_{ox} + \varepsilon_s A/\lambda} \tag{1}$$

where $\varepsilon_s = 9.5$ is the dielectric constant of GaN, $\lambda = (\varepsilon_s\varepsilon_0 k_B T/q^2 N_D)^{1/2}$ is the Debye length of n-GaN [10], T is the absolute temperature, q is the electron charge, k_B is the Boltzmann constant, and $N_D = 6 \times 10^{17}$ cm^{-3} is the electron concentration of n-GaN. The relative shifts in V_{FB} during the sweep down (10 to -10 V) and up (-10 to 10 V) for the surface treatments of GaN with various treatments are consistent with the presence of interface state trap densities at or near the GaN/Al$_2$O$_3$

interface, which was based on the C−V curve (Figure 2). The hysteresis at V_{FB} (shift in V_{FB}) was used to approximate the interface state trap densities in each sample, according to the C–V characteristics. The flat-band voltage V_{FB} of the sample is shown in Table 2, and the threshold onset voltage V_{th} is obtained using [11]:

$$V_{th} = V_{FB} - 2|\phi_b| - \frac{\sqrt{4q\varepsilon_s\varepsilon_o N_D |\phi_b|}}{\varepsilon_{ox}\varepsilon_0/t_{ox}} \qquad (2)$$

where $\phi_b = (k_B T/q)\ln(N_D/n_i)$, $n_i = 2.0 \times 10^{-10}$ cm^{-3} is the intrinsic carrier concentration of GaN at room temperature [12], $\varepsilon_{ox} = 9.9$ is the dielectric constant of Al$_2$O$_3$, and t_{ox} is the thickness of the Al$_2$O$_3$ dielectric. The relative shift (ΔV_{th}) is the variation in V_{FB} during the sweep down (10 to −10 V) and up (−10 to 10 V) and is calculated using Equation (2) for the surface treatment of GaN with different treatments. A small voltage shift (ΔV_{th}) is attributable to the different charging conditions of the interface states with different chemical treatments. The interface state trap densities (D_{it}) can be estimated using [13,14]:

$$D_{it} = \frac{C_{ox}\Delta V_{FB}(T)}{q} \qquad (3)$$

where q and C_{ox} are the electron charge and accumulation capacitance per unit of area, respectively. In the worst case scenario, treatment 1 showed a D_{it} value of 1.74×10^{12} cm^{-2}. The lowest D_{it}, 8.30×10^{11} cm^{-2}, was obtained using treatment 4. The D_{it} of GaN treated with treatment 4 was reduced by approximately 50% compared with that of GaN treated with treatment 1. Treatment 4 consisted of HF, HCl, and NH$_4$OH, which is used for removing native oxides, metal ions, and organic contaminants of GaN. Therefore, the GaN surface treated with treatment 4 showed the cleanest surface, and the lowest D_{it} was obtained. The oxide capacitance (capacitance at accumulation, C_{ox}), flat band voltage (V_{FB}), real thickness of Al$_2$O$_3$ measured using transmission electron microscopy, flat band capacitance (C_{FB}), and interface state trap densities (D_{it}) are summarized in Table 2. The different chemical treatment causes a different D_{it} value, the chemical treatment not only affects the D_{it}, but also modified the GaN surface contact properties with Ti/Al/Ti/Au. Therefore, D_{it} value was related to surface contact resistance with Ti/Al/Ti/Au. The high D_{it} value will result in the contact resistance. Because the D_{it} value is sensitive to the GaN surface condition with different chemical treatment, in order to further understand the surface condition after chemical treatment, the XPS spectra analysis and XPS Ga-O/Ga-N ratio were used to explain the D_{it} value changed with different treatment recipes.

The chemical treatment technology modified the n-GaN surface, subsequent dielectrics material growth, contact resistance with metal material, and HEMT device performance. A circular transmission line model was used to evaluate the ρ_c of GaN contacted with Ti/Al/Ti/Au. The results indicated that the ρ_c of the samples treated with treatments 1–4 were 2.77×10^{-4}, 3.51×10^{-4}, 2.63×10^{-4}, and 2.20×10^{-4} Ω-cm^2, respectively. The different ρ_c value was related to the GaN surface barrier height, the details were described in Figure 5. The ρ_c of GaN treated with treatment 4 was reduced by approximately 22% compared with that of GaN treated with treatment 1. The contact characteristics of GaN with Ti/Al/Ti/Au were affected by the coverage of oxide and carbon contaminants. Complex cleaning agents, such as treatments 2–4, were used to remove or reduce the contaminants. Treatments 2–4 contained HCl and HF, which are known to remove oxides from Ga-based semiconductors [15]. However, HCl and HF wet-chemical pretreatments are more effective in producing the lowest coverage of oxide and carbon contaminants [16] to modify the contact characteristic of GaN with Ti/Al/Ti/Au. Treatment 4 consisted of an alkaline solution, acidic solution, and diluted HF, and was used to remove organic contaminants, metal ions, and native oxides. Notably, NH$_4$OH:H$_2$O (1:3) predominantly removes gallium oxide (Ga$_2$O$_3$) from the GaN surface [17] and organic contaminants, thus improving the adhesion ability of the metal film. Therefore, treatment 4 resulted in the lowest contamination and cleanest surface; thus, the lowest ρ_c was obtained.

Table 2. The C_{ox}, V_{FB}, hysteresis at V_{FB}, and D_{it} of an MOS capacitor after different chemical treatments and the specific contact resistance (ρ_c) of GaN contacted with Ti/Al/Ti/Au.

Item	Treatment 1	Treatment 2	Treatment 3	Treatment 4
C_{ox} (pF)	15.6	15.1	14.9	14.9
C_{FB} (pF)	13.98	13.57	13.41	13.41
Hysteresis at V_{FB} (V)	1.4	1.3	0.9	0.7
t_{ox} (nm) (Thickness observed by TEM)	49.02	48.50	47.13	48.16
D_{it} (cm^{-2})	1.74×10^{11}	1.56×10^{12}	1.07×10^{12}	8.30×10^{11}
ρ_c of CTLM (Ω-cm^2)	2.77×10^4	3.51×10^4	2.63×10^4	2.20×10^4

The ρ_c is related to the surface barrier height of the GaN surface. The Ga–O to Ga–N ratio of the Ga3d peak was used to facilitate the analysis of the surface barrier height of GaN through various chemical treatments. XPS was used to study the surface composition on the GaN surface by using different chemical treatments. Figure 3a–d shows the Ga3d core level of the XPS spectra. The Ga3d peaks of GaN obtained using treatments 1–4 appeared at 19.4, 19.4, 19.6, and 19.4 eV, respectively. A blue shift of approximately 0.2 eV was observed toward the high binding energy in a sample treated with treatment 3, compared with samples treated with the other treatments. This type of shift is assumed to have been caused by the loss of N at the surface or the creation of N vacancies, which would increase the n-type doping at the surface [18]. In addition, the shift could have been caused by the BCl$_3$ plasma. Moreover, the Ga3d of XPS spectra photoelectrons can be separated into Ga–O and Ga–N components for various treatments. The main peak at a binding energy of 19.3 eV corresponded to the Ga–N bond, and the second peak at 20.3 eV corresponded to the Ga–O bond, thus confirming the presence of Ga$_2$O$_3$ as the native oxide layer on top of the GaN layer. The intensity and area of the Ga–O core level of the Ga3d peak after different treatments are functions of surface conditions. The Ga–O core level is reduced by a more complex chemical treatment association. The reduction of the Ga–O core level indicated that the Ga$_2$O$_3$ layer was effectively removed or reduced. Otherwise, the remaining Ga$_2$O$_3$ layer might affect the quality of the ohmic contact and thus increase the contact resistance between GaN and metal.

Figure 4 shows the integrated Ga–O core level levels, normalized using the Ga–N core level as a function of surface conditions to evaluate the residual native oxide layer on the GaN surface. The lowest Ga–O to Ga–N ratio of the Ga3d peak was obtained after treatment with treatment 4. Therefore, the ρ_c of GaN contacted with Ti/Al/Ti/Au was the lowest for the sample treated with treatment 4. The area ratios of Ga–O to Ga–N obtained after different treatments are consistent with the contact characteristics of GaN with Ti/Al/Ti/Au (Table 2). In addition, the peak ratio of Ga–O to Ga–N obtained using treatment 2 was the highest. This is because the Ga–O area of Ga3d obtained using treatment 2 was the highest, resulting in a high contact resistance. The Ga–O area was the highest because the fresh dangling bonds created by O$_2$ plasma and trap more O$_2$ existing in HCl:H$_2$O and HF:H$_2$O solutions. The [Ga–O]/[Ga–N] ratios decreased because of the reduction in the O concentration and/or increase in the N concentration at the GaN surface [2]. In our study, samples treated with treatment 4 had the lowest [Ga–O]/[Ga–N] ratio. After O$_2$ plasma treatment, the residual GaO could be further etched by HF:H$_2$O, following oxidation by NH$_4$OH:H$_2$O and etched away by HCl:H$_2$O and HF:H$_2$O. The [Ga–O] decreased and [Ga–N] increased during GaO etching.

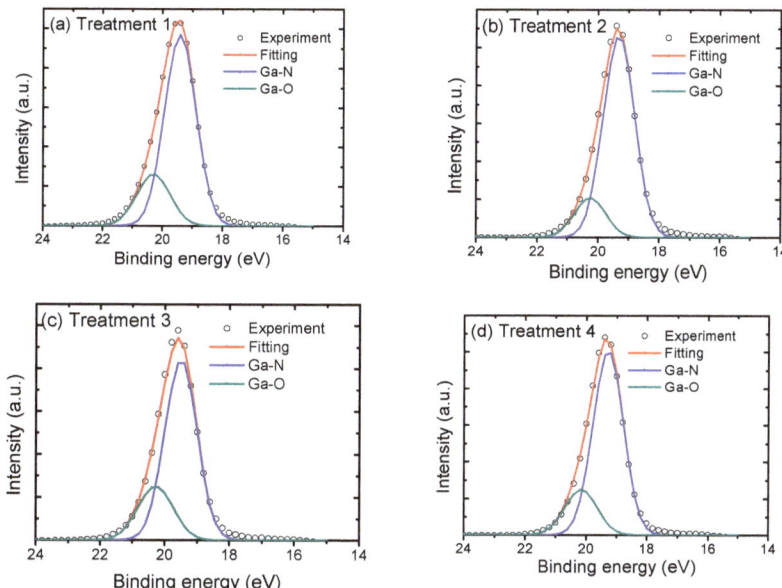

Figure 3. XPS spectra of the Ga3d core levels of the GaN layer after treatments with (**a**) treatment 1, (**b**) treatment 2, (**c**) treatment 3, and (**d**) treatment 4.

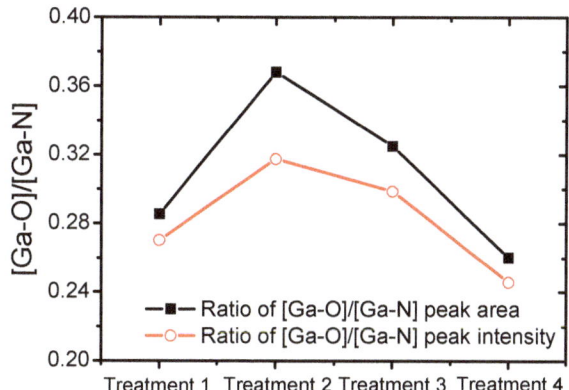

Figure 4. Ratio of Ga–O/Ga–N obtained using the XPS spectra from the Ga3d core levels of the GaN layer.

Figure 5 shows the XPS valence band spectra of GaN treated with various treatments. A binding energy of 0 eV on the horizontal axis corresponded to the energy position of the Fermi level (E_f) at the surface. The energy position of the valence band maximum (VBM) was determined by linearly extrapolating the spectrum near the onset [19] to calculate the surface barrier height of GaN. The surface barrier height (Φ_B) is defined as $E_c - E_f$, where E_c is the energy position of the conduction band minimum. The VBM of GaN treated with different treatments was lower than the Fermi level by approximately 1.68–2.26 eV.

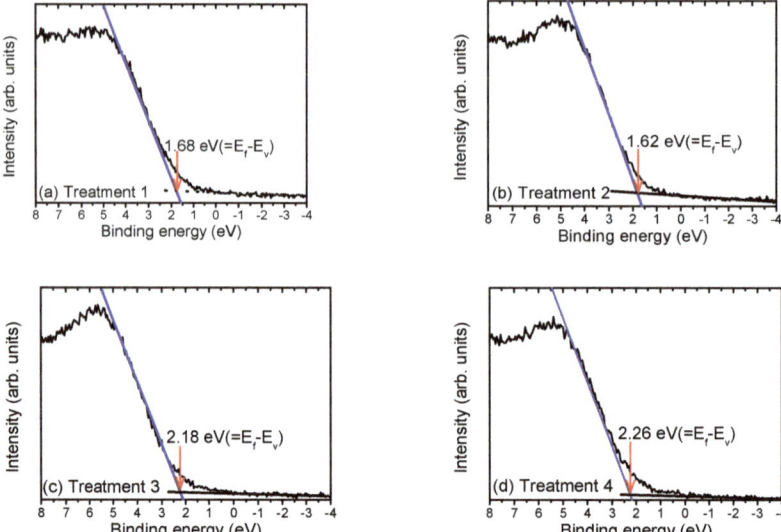

Figure 5. XPS valence band spectra of GaN after treatments with (**a**) treatment 1, (**b**) treatment 2, (**c**) treatment 3, and (**d**) treatment 4.

The surface treatment of GaN modifies the GaN surface condition, including surface barrier height, binding energy and surface quality, which will further change the GaN ohmic contact properties with Ti/Al/Ti/Au. The XPS valence band and XPS spectra were powerful enough to determine the surface barrier height of GaN after treatment. The high surface barrier height caused the poor ohmic contact. The surface barrier height of GaN after treatment with treatment 1–4 is shown in Figure 6. The trend of the surface barrier height is consistent with that of the ρ_c. A high ρ_c indicates a high surface barrier height of GaN. The surface barrier height of the GaN surface after BCl_3 plasma treatment (in treatment 3) was lower than that of GaN after the treatment 1 and 2 treatments. The BCl_3 plasma increased the surface N vacancy of GaN, which acts as a donor-type density for electrons, thus increasing the surface doping concentration [20]. The increase in the donor-type density of the GaN surface further improved the metallurgical process to reduce the surface barrier height of GaN during RTA. During RTA, Ti undergoes a metallurgical reaction with GaN, forming interfacial nitrides such as TiN. This can cause the GaN subsurface below the TiN to be heavily doped (n-type) [21]. Therefore, the surface barrier height of GaN after treatment 3 was lower than that after treatments 1 and 2, and the surface barrier height of GaN after treatment 4 was the lowest. This result indicates that treatment 4 had the highest surface donor density, which contributed to a reduction in the ohmic contact resistance. A lower surface barrier height of GaN indicated a low contact resistance of GaN with Ti/Al/Ti/Au, thereby easily forming an ohmic contact for device applications.

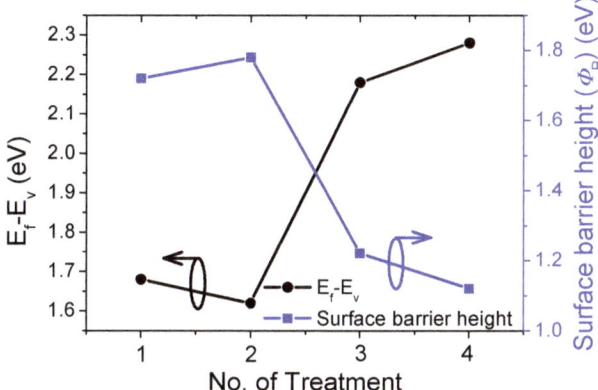

Figure 6. $E_f - E_v$ and surface barrier heights of GaN after treatments with treatment 1–4.

4. Conclusions

In summary, chemical treatments were successfully used to reduce the D_{it} and improve the surface quality. The lowest D_{it} of 8.30×10^{11} cm^{-2} of the MOS capacitor was obtained when the sample was treated with treatment 4 because treatment 4 consisted of an alkaline solution, acidic solution, and diluted HCl, and was used to remove organic contaminants, metal ions, and native oxides. The chemical solution treatment not only reduced the contamination but also introduced the donor density to change the ohmic contact property of n-GaN with metal material.

Author Contributions: R.H. and D.W.; methodology, R.H., M.T. and D.W.; formal analysis, M.T.; investigation, D.W.; data curation, M.T.; writing—original draft preparation, R.H.; writing—review and editing, R.H. and D.W.; supervision, R.H.; project administration.

Funding: This research was funded by Ministry of Science and Technology under contract numbers MOST 107-2221-E-009-117-MY3, 107-2262-E-009-018-CC2, MOST107-3017-F009-003 and 105-2221-E-009-183-MY3.

Acknowledgments: This study was financially supported by the Ministry of Science and Technology under contract numbers MOST 107-2221-E-009-117-MY3, 107-2262-E-009-018-CC2 and 105-2221-E-009-183-MY3.

Conflicts of Interest: The authors declare no conflict of interest.

References

1. Kern, W. The Evolution of Silicon Wafer Cleaning Technology. *J. Electrochem. Soc.* **1990**, *137*, 1887–1990. [CrossRef]
2. Selvanathan, D.; Mohammed, F.M.; Bae, J.O.; Adesida, I.; Bogart, K.H.A. Investigation of surface treatment schemes on n-type GaN and Al$_{0.20}$Ga$_{0.80}$N. *J. Vac. Sci. Technol. B* **2005**, *23*, 2538–2544. [CrossRef]
3. Cao, X.A.; Piao, H.; LeBoeuf, S.F.; Li, J.; Lin, J.Y.; Jiang, H.X. Effects of plasma treatment on the Ohmic characteristics of Ti/Al/Ti/Au contacts to n-AlGaN. *Appl. Phys. Lett.* **2006**, *89*, 082109. [CrossRef]
4. Hori, Y.; Yatabe, Z.; Hashizume, T. Characterization of interface states in Al$_2$O$_3$/AlGaN/GaN structures for improved performance of high-electron-mobility transistors. *J. Appl. Phys.* **2013**, *114*, 244503. [CrossRef]
5. Vetury, R.; Zhang, N.Q.; Keller, S.; Mishra Umesh, K. The Impact of Surface States on the DC and RF Characteristics of AlGaN/GaN HFETs. *IEEE Trans. Electron. Devices* **2001**, *48*, 560–566. [CrossRef]
6. Roccaforte, F.; Fiorenza, P.; Greco, G.; Vivona, M.; Nigro, R.L.; Giannazzo, F.; Patti, A.; Saggio, M. Recent advances on dielectrics technology for SiCand GaN power devices. *Appl. Surf. Sci.* **2014**, *301*, 9–18. [CrossRef]
7. Hashizume, T.; Nishiguchi, K.; Kaneki, S.; Kuzmik, J.; Yatabe, Z. State of the art on gate insulation and surface passivation for GaN-based power HEMTs. *Mater. Sci. Semicond. Proc.* **2018**, *78*, 85–95. [CrossRef]
8. Schilirò, E.; Fiorenza, P.; Greco, G.; Roccaforte, F.; Nigro, R.L. Plasma enhanced atomic layer deposition of Al$_2$O$_3$ gate dielectric thin films on AlGaN/GaN substrates: The role of surface predeposition treatments. *J. Vac. Sci. Technol. A* **2017**, *35*, 01B140. [CrossRef]

9. Nicollian, E.H.; Brews, J.R. *MOS (Metal Oxide Semiconductor) Physics and Technology*; Wiley Publishers: New York, NY, USA, 1982; pp. 462–463. ISBN 978-0-471-43079-7.
10. Razeghi, M.; Rogalski, A. Semiconductor ultraviolet detectors. *J. Appl. Phys.* **1996**, *79*, 7433–7473. [CrossRef]
11. Muller, R.S.; Kamins, T.I. *Device Electronics for Integrated Circuits*, 2nd ed.; Wiley Publishers: New York, NY, USA, 1986; pp. 50–54, ISBN 9780471593980.
12. Casey, H.C., Jr.; Fountain, G.G.; Alley, R.G.; Keller, B.P.; DenBaars, S.P. Low interface trap density for remote plasma deposited SiO_2 on n-type GaN. *Appl. Phys. Lett.* **1996**, *68*, 1850–1852. [CrossRef]
13. Fiorenza, P.; Greco, G.; Schilirò, E.; Iucolano, F.; Nigro, R.L.; Roccaforte, F. Determining oxide trapped charges in Al_2O_3 insulating films on recessed AlGaN/GaN heterostructures by gate capacitance transients measurements. *Jpn. J. Appl. Phys.* **2018**, *57*, 050307. [CrossRef]
14. Schilirò, E.; Nigro, R.L.; Fiorenza, P.; Roccaforte, F. Negative charge trapping effects in Al_2O_3 films grown by atomic layer deposition onto thermally oxidized 4H-SiC. *AIP Adv.* **2016**, *6*, 075021. [CrossRef]
15. Wilmsen, C.W. *Physics and Chemistry of III-V Compound Semiconductor Interfaces*; Springer Publishers: New York, NY, USA; London, UK, 1985; pp. 1–72. ISBN 978-1-4684-4835-1.
16. Smith, L.L.; King, S.W.; Nemanich, R.J.; Davis, R.F. Cleaning of GaN surfaces. *J. Electron. Mater.* **1996**, *25*, 805–810. [CrossRef]
17. Prabhakaran, K.; Andersson, T.G.; Nozawa, K. Nature of native oxide on GaN surface and its reaction with Al. *Appl. Phys. Lett.* **1996**, *69*, 3213–3214. [CrossRef]
18. Ping, A.T.; Chen, Q.; Yang, J.W.; Khan, M.A.; Adesida, I. The effects of reactive ion etching-induced damage on the characteristics of ohmic contacts to n-Type GaN. *J. Electron. Mater.* **1998**, *27*, 261–265. [CrossRef]
19. Higashiwaki, M.; Chowdhury, S.; Swenson, B.L.; Mishra, U.K. Effects of oxidation on surface chemical states and barrier height of AlGaN/GaN heterostructures. *Appl. Phys. Lett.* **2010**, *97*, 222104. [CrossRef]
20. Fujishima, T.; Joglekar, S.; Piedra, D.; Lee, H.S.; Zhang, Y.; Uedono, A.; Palacios, T. Formation of low resistance ohmic contacts in GaN-based high electron mobility transistors with BCl_3 surface plasma treatment. *Appl. Phys. Lett.* **2009**, *103*, 083508. [CrossRef]
21. Kim, J.K.; Jang, H.W.; Lee, J.L. Mechanism for Ohmic contact formation of Ti on n-type GaN investigated using synchrotron radiation photoemission spectroscopy. *J. Appl. Phys.* **2002**, *91*, 9214–9217. [CrossRef]

© 2018 by the authors. Licensee MDPI, Basel, Switzerland. This article is an open access article distributed under the terms and conditions of the Creative Commons Attribution (CC BY) license (http://creativecommons.org/licenses/by/4.0/).

Article

Role of Hydrogen in Active Layer of Oxide-Semiconductor-Based Thin Film Transistors

Hee Yeon Noh, Joonwoo Kim, June-Seo Kim, Myoung-Jae Lee and Hyeon-Jun Lee *

Intelligent Devices & Systems Research Group, Institute of Convergence, DGIST, Daegu 42988, Korea; heeyeon@dgist.ac.kr (H.Y.N.); jwkim1206@dgist.ac.kr (J.K.); spin2mtj@dgist.ac.kr (J.-S.K.); myoungjae.lee@dgist.ac.kr (M.-J.L.)
* Correspondence: dear.hjlee@dgist.ac.kr; Tel.: +82-53-785-3503

Received: 26 December 2018; Accepted: 30 January 2019; Published: 31 January 2019

Abstract: Hydrogen in oxide systems plays a very important role in determining the major physical characteristics of such systems. In this study, we investigated the effect of hydrogen in oxide host systems for various oxygen environments that acted as amorphous oxide semiconductors. The oxygen environment in the sample was controlled by the oxygen gas partial pressure in the radio-frequency-sputtering process. It was confirmed that the hydrogen introduced by the passivation layer not only acted as a "killer" of oxygen deficiencies but also as the "creator" of the defects depending on the density of oxide states. Even if hydrogen is not injected, its role can change owing to unintentionally injected hydrogen, which leads to conflicting results. We discuss herein the correlation with hydrogen in the oxide semiconductor with excess or lack of oxygen through device simulation and elemental analysis.

Keywords: oxide semiconductor; InGaZnOx; hydrogen; oxygen deficiency; technology computer aided design (TCAD)

1. Introduction

Amorphous oxide semiconductors have been extensively investigated for application in transparent electronics, backplanes of large-area active-organic light-emitting diode displays, and liquid crystal displays [1]. Among devices made using various oxide semiconductor materials, amorphous indium–gallium–zinc–oxide (a-IGZO) thin film transistors (TFTs) exhibit a wide range of electron densities (N_e) ranging [2] from 10^{11} to 10^{19} cm^{-3}. Oxide semiconductors with wide band gaps, along with these high electron densities, are considered to be a promising alternative to amorphous silicon TFTs because they can provide high mobility and low leakage currents.

However, a-IGZO TFTs still present problems in the form of various defects between the metal and the oxygen, such as deficiencies (similar to vacancies in a crystal system) and interstitial defects, which they strongly affect the electrical properties of the devices and their reliability in practical applications [3–6]. In addition to such intrinsic defects, it is well known that hydrogen as an impurity can also affect the electrical properties of a-IGZO TFTs. Hydrogen, which results in high conductivity due to form the donor level above conduction band, can be found in various forms in crystalline oxide semiconductors [7,8]. It can be present as interstitial hydrogen (H_i), located at bond-centered or antibonding sites owing to O-H bonds [9]. A second type can be trapped at oxygen deficiencies (H_o) [10]. The final kind is unbound hydrogen, or the interstitial H_2 [11]. Hydrogen impurity present in high densities of 10^{20}–10^{21} cm^{-3} is always found in a-IGZO films [2]. When it forms bonds with oxygen and generates negatively charged hydroxide anions, hydrogen becomes a shallow donor, generating free carriers in the oxide semiconductors. On the other hand, hydrogen can deactivate defects in a-IGZO due to compensation by excess oxygen, which manifests as defect passivation [12]. Many studies have been conducted in an effort to reveal the effects of hydrogen on defects of a-IGZO and have involved techniques such as hydrogen plasma treatment [13], annealing temperature control in hydrogen ambient [14], and hydrogen injection through buffer layers [15]. Oh et al. investigated the hydrogenation of a-IGZO films under high-pressure hydrogen annealing at different temperatures [16]. A very high carrier density (N_e) of 1.63×10^{19} cm^{-3} was observed as a result, which is 10^5 times higher than that of as-grown films. Nguyen et al. reported the influence of hydrogen arising from silane (SiH$_4$) in the passivation layer on the device properties [17]. It was clearly observed that the V_{th} shifted negatively and the leakage current seemed to increase on increasing the silane concentration. This hydrogen incorporation into a-IGZO induced excess carriers in channels, making the system a better conductor. Han et al. studied the effect of a multi-layered buffer on the electrical properties of a-IGZO TFTs [18]. When the amount of diffused hydrogen in the a-IGZO layer was under a critical value, electron trapping at the a-IGZO/insulator interface was effectively suppressed by deactivation of defect states in the a-IGZO layer. Conversely, hydrogen diffusion from the hydrogen-rich layer caused a conducting channel. Other roles were assumed for hydrogen in a-IGZO TFTs and difficulty in achieving precise control of hydrogen during the fabrication of TFTs make this issue more difficult to resolve [18].

So far, studies in this area have been mainly conducted focusing on hydrogen, regardless of the oxide semiconductor host system. In this work, we studied the effect of hydrogen in host systems with various oxygen environments in amorphous oxide semiconductors. We found that the hydrogen introduced by the passivation layer not only acts as the "killer" of oxygen deficiencies but also as the "creator" of defect levels depending on the density of oxide states. Its role strongly depends on the host oxide semiconductor system, especially on the oxygen environments between the metal cations. Herein, we have discussed the role of hydrogen in oxygen-excess systems as well as oxygen-deficient systems, and have confirmed our results through device simulation and elemental analysis.

2. Materials and Methods

The TFTs with a bottom-gate inverted staggered structure, as shown in Figure 1a, were fabricated on a silicon wafer substrate. A Au (100 nm) gate electrode was deposited by the direct current (DC) magnetron sputtering method. Dual layers of a-SiN$_x$ (50 nm)/SiO$_x$ (150 nm) were subsequently deposited as a gate insulator by plasma-enhanced chemical vapor deposition (PECVD, KCT, Inc., Gyeonggi-do, Korea) at 350 °C. A 40-nm-thick a-IGZO active layer (In:Ga:Zn = 1:1:1 by mol%) was deposited by radio frequency (RF) magnetron sputtering (4 inch target, distance of 120 mm between the source and the substrate, Korea Vacuum Tech, Inc., Gyeonggi-do, Korea) at 100 °C using a gas mixture of Ar and O$_2$ and an input power of 200 W (5 mTorr). A Au S/D layer (100 nm) was deposited and patterned by photolithography (W/L = 10 µm/10 µm) followed by lift-off. The gate-via was formed by a reactive ion dry etching process. The single passivation layer (without SiO$_x$N$_y$) of SiO$_x$ (200 nm) and dual passivation layers of SiO$_x$ (100 nm)/ SiO$_x$N$_y$ (100 nm) were subsequently deposited under the same conditions with the gate insulator layers. Finally, the deposited layers were annealed at 350 °C for 1 h under vacuum. SiO$_x$N$_y$ was considered to be a likely source of hydrogen and its value was controlled by the ratio of the precursor materials, the nitrous oxide (N$_2$O) and ammonia (NH$_3$), at values of SiO$_x$, SiO$_{1.5}$N$_{0.01}$, and SiO$_{1.45}$N$_{0.02}$ (The ratio of oxygen and nitrogen was estimated by x-ray photoelectron spectroscopy (XPS)). Samples were prepared using RF-sputtering under three different oxygen partial pressures. Detailed conditions of which has been provided in Table 1. A Keithley 2636B source meter (Keithley Instrument, Inc., OH, USA) was used to determine the electrical characteristics. The surface properties and bonding character were investigated by XPS (Thermo Scientific, Inc., MA, USA) and time-of-flight secondary ion mass spectroscopy (TOF-SIMS, Ion-tof, Inc., Münster, Germany) was utilized to measure and analyze the hydrogen concentration in the active layer. The device was simulated using technology computer aided design (TCAD, Silvaco, Inc., CA, USA, 2018) to understand the electron transport properties, using Silvaco's 2D ATLAS simulator package (Silvaco Inc., CA, USA, 2018) [19,20].

Table 1. Ar/O_2 partial pressure conditions in the radio frequency (RF) sputtering.

Device	Ar (sccm)	O_2 (sccm)
Device A	50	0.7
Device B	50	1.0
Device C	50	5.0

3. Results

To identify the effect of hydrogen for different host system environments in the active layer, hydrogen was introduced by various SiO_xN_y passivation layer conditions in the *a*-IGZO. The hydrogen densities depending on the passivation in *a*-IGZO are shown in Figure 1b. The density of hydrogen was estimated to be >10^{20} atom/cm^{-3} in the entire samples and the density increased with increasing nitrogen concentration in the passivation layer. Since the nitrogen precursor of the SiO_xN_y layer was NH_3 in the PECVD process, the hydrogen density strongly depended on the nitrogen concentration [17]. Figure 2 compares the transfer characteristics (V_{ds} = 10 V) of *a*-IGZO TFTs influenced by hydrogen via the passivation layer depending on the oxygen partial pressure, "Device A, Device B, and Device C".

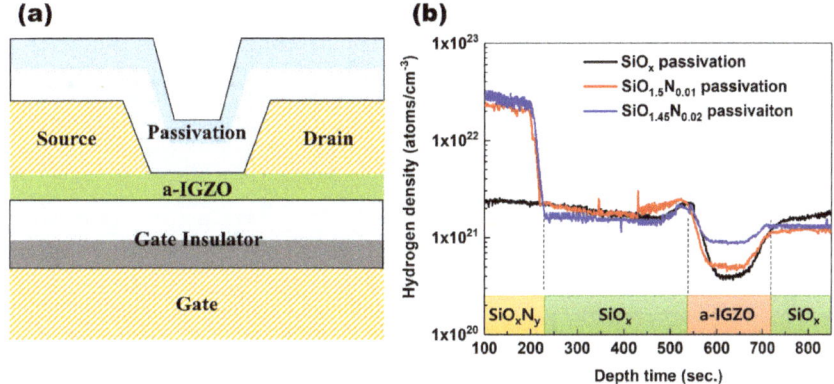

Figure 1. (a) Cross-sectional schematic of the bottom-gate inverted staggered structure of a-InGaZnO TFT. (b) TOF-SIMS depth profile of hydrogen atoms in a-IGZO of device B.

An increasing threshold voltage (the linear extraction method in the saturation region) was observed as the oxygen partial pressure increased, as seen in Figure 2a. Increase in the oxygen partial pressure from 0.7 to 1 sccm gave rise to a decrease in the subthreshold swing (voltage with an order current at the maximum I-V gradient; ln 10 $dV_G/d(\ln I_D)$) value from 0.58 to 0.36 V/dec. However, the partial pressure continuously increasing to 5 sccm caused an increase in the subthreshold swing value. It seemed that the oxygen partial pressure alone was able to not only control the oxygen deficiency in the oxide semiconductor devices but also create an impurity level due to the excess oxygen. Injection of excessive oxygen causes an increase in subthreshold swing due to an increasing number of defect states and the positive shift in the threshold voltage, as shown in Figure 2a. The electrical properties changing as a function of the additional SiO_xN_y layer is shown in Figure 2b,c. At a low oxygen partial pressure (Device A), the threshold voltage, shown in Figure 2a, increased to 3.34 V but the subthreshold swing value decreased to 0.50 V/dec when the second passivation $SiO_{1.5}N_{0.01}$ layer was deposited as shown in Figure 2d,e. On the other hand, at a high oxygen partial pressure (Device C), we observed decreasing threshold voltage and subthreshold swing values. It was proposed that two different phenomena were brought about by the additional hydrogen in the two different host systems. The first was a passivation effect, which is a well-known phenomenon that reduces the oxygen deficiency states via hydrogen substitution in oxygen-deficient sites. In the case of the excess-oxygen sample, it seemed that the hydrogen reduced the defect states that were below the conduction band minimum and were formed by the excess oxygen in the oxide semiconductor. Above all, it was interesting to observe that the additional hydrogen could form defect states. The sample showing good electrical transfer characteristics, Device B in Figure 2a, underwent drastic changes in the electric characteristics through hydrogen injection via added passivation by $SiO_{1.5}N_{0.01}$ as shown in Figure 2b. Figure 2c shows the electrical transfer curves of the sample with a $SiO_{1.45}N_{0.02}$ passivation layer and a high subthreshold swing value of around 1.0 V/dec. This suggested that the continued increasing amount of hydrogen gave rise dominant formation of defect states.

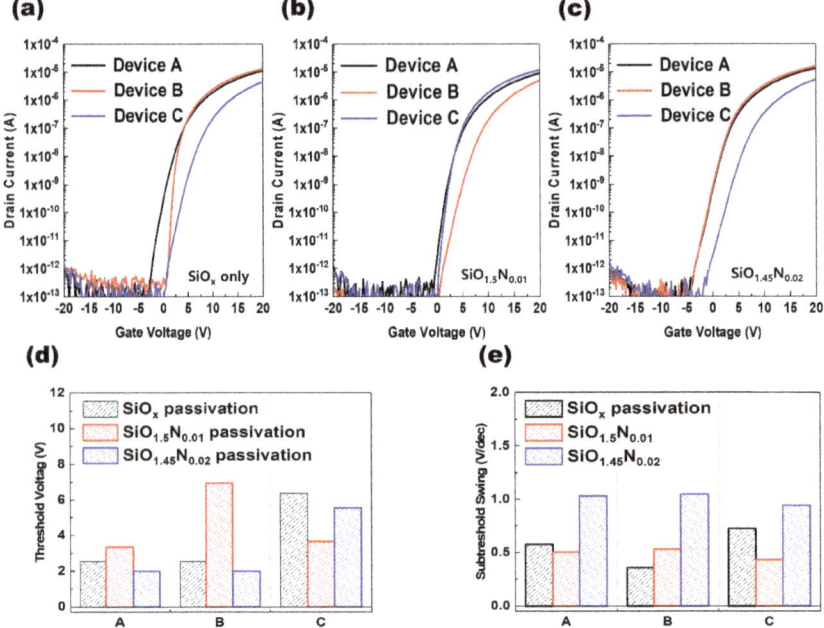

Figure 2. Electric transfer curves of *a*-IGZO TFTs with passivation by (**a**) SiO_x, (**b**) $SiO_{1.5}N_{0.01}$, and (**c**) $SiO_{1.45}N_{0.02}$. Summary of the (**d**) threshold voltage and (**e**) subthreshold swing.

In order to understand the defect generation by hydrogen through the passivation layer, a 2-dimensional numerical TCAD simulation was used to simulate the device characteristics. In Figure 3a, both black (V_{ds} 0.1 V) and red (V_{ds} 10 V) solid lines show the measured I–V transfer curves of "Device B" with SiO_x passivation. Both dashed lines indicate the simulated I–V transfer curves and these matched well with the measured I–V data. Figure 3b shows the initial band gap states along with the tailing and sub-gap states of the *a*-IGZO. The peak density of the donor-like tail state near the valence band and the donor-like Gaussian state below the conduction band were calculated to be 2.0×10^{19} cm^{-3}eV^{-1} (width of the state: 0.2 eV) and 3.2×10^{19} cm^{-3}eV^{-1} (width of the state: 0.037 eV), respectively. A broad acceptor-like Gaussian state, 0.9 eV, and a narrow donor-like Gaussian state, 0.1 eV, were confirmed on the band gap. A decreasing oxygen partial pressure reduced the acceptor-like Gaussian broadening but increased the width of the donor-like Gaussian state as well as that of the donor-like tailing state, as shown in Figure 3c. In the excess-oxygen case, broadening of the acceptor-like Gaussian state increased significantly from 0.61 eV to 1.05 eV, as shown in Figure 3d. Figure 3e,f show the simulated density of states for the effects of the hydrogen. The increasing hydrogen concentration in the active layer owing to the passivation layer ($SiO_{0.5}N_{0.01}$) increased the defect states, especially the donor-like tailing states and acceptor-like Gaussian states. Figure 3e shows the density of state in Device B with $SiO_{0.5}N_{0.01}$ passivation, which was comparable to that in Device B with SiO_x passivation, as shown in Figure 3b. In addition, the hydrogen introduced in the oxygen-deficient sample, Device A with SiO_x passivation, reduced the defect energy levels. In this case, the width of the donor-like tailing state and donor-like Gaussian states decreased drastically and the band edge below the conduction band minimum was relatively sharp, as shown in Figure 3f.

Hydrogen is a typical element that is unintentionally injected into sample layers. Because this hydrogen has a significant effect on the device properties, hydrogen can be used in various ways in its semiconductor process. Hydrogen in InGaZnO, a major oxide semiconductor, is able to replace [21] oxygen deficiency sites, and the defect states arising due to a lack of oxygen are removed as shown in Figure 4a. Due to these passivation effects of the oxygen deficiency state, many studies have used hydrogenation processes to secure initial electrical characteristics and long-time reliability. However, this hydrogen injection cannot only play a positive role. The abovementioned results depicted in Figures 2 and 3 confirmed that excessive injection of hydrogen can cause various defect states. It was confirmed by 2-dimensional TCAD device simulation that excessive hydrogen can give rise to donor-like tailing states and acceptor-like Gaussian defect states. Figure 4b,c are schematics of the possible impurities of hydrogen injection. Injection of excess hydrogen can lead to its conversion to the molecular H_2 state [11] (Figure 4b), and the reduction reaction via hydrogenation can also weaken the binding energy of oxygen and weakly bound In or Zn, as shown in Figure 4c.

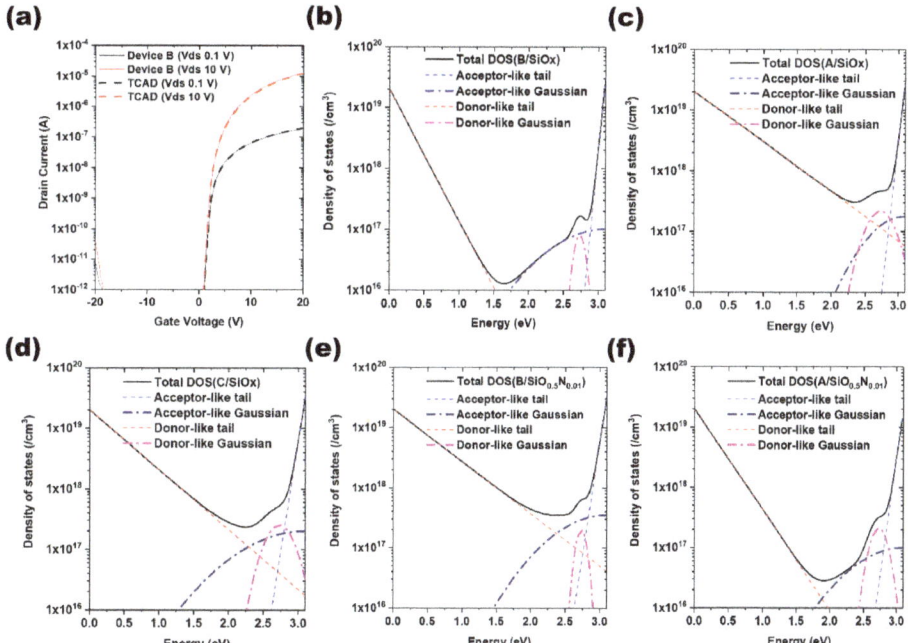

Figure 3. (**a**) I–V transfer curves for Device B with SiO_x passivation, measured and simulated. TCAD simulated density of states for (**b**) Device B, with SiO_x; (**c**) Device A, with SiO_x; (**d**) Device C, with SiO_x; (**e**) Device B, with $SiO_{0.5}N_{0.01}$; and (**f**) Device A, with $SiO_{0.5}N_{0.01}$. The variation of the oxygen partial pressure give rise to noticeable change the "Donor-like Gaussian" state as well as "Donor-like tail" state. The center of the "Donor-like Gaussian" state is around 2.73 eV.

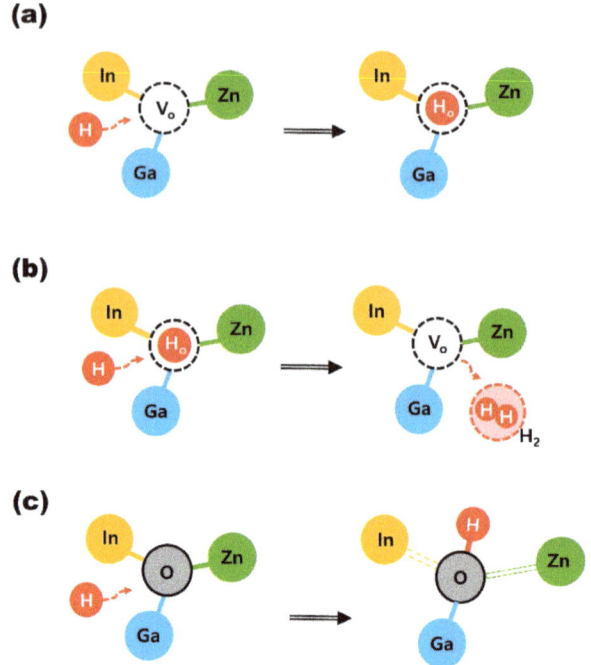

Figure 4. Schematic structure for the hydrogen and oxygen migration process for cases involving (**a**) trapped hydrogen in oxygen-deficient state, (**b**) unbound hydrogen, and (**c**) weak bonding of In and Zn with oxygen bound to hydrogen.

4. Conclusions

In this paper, we reported the effects of hydrogen on the host system for various oxygen environments in amorphous oxide semiconductors. We employed a-IGZO as an oxide semiconductor for our investigation. Hydrogen injection was adjusted by changing the passivation layer conditions and increased hydrogen concentration was confirmed using SIMS. Electrical measurements and simulations showed that hydrogen played a role in eliminating oxygen deficiencies, which is the biggest issue in oxide semiconductors, and excess injection of hydrogen produced additional structural defects in the host system. On injecting hydrogen into oxide semiconductors, in particular, hydrogen may play different roles as the species that reduces defect states depending on the oxygen environments of the host system or the one that increases defect states.

Author Contributions: H.-J.L. designed the experimental concept and wrote the main text of the manuscript. H.Y.N. fabricated the device. J.K., J.-S.K. and M.-J.L. contributed to the circuit design and the overall understanding of the device physics. All authors reviewed the manuscript.

Funding: This research was supported in part by the DGIST R&D Program of the Ministry of Science, ICT, and Future planning (18-NT-01) and in part by the Basic Science Research Program through the National Research Foundation of Korea (NRF) under Grant NRF-2018R1D1A1B07041075.

Conflicts of Interest: The authors declare no conflict of interest.

References

1. Nomura, K.; Ohta, H.; Takagi, A.; Kamiya, T.; Hirano, M.; Hosono, H. Room-temperature fabrication of transparent flexible thin-film transistors using amorphous oxide semiconductors. *Nature* **2004**, *432*, 488–492. [CrossRef] [PubMed]

2. Tang, H.; Kishida, Y.; Ide, K.; Toda, Y.; Hiramatsu, H.; Matsuishi, S.; Ueda, S.; Ohashi, N.; Kumomi, H.; Hosono, H.; et al. Multiple Roles of Hydrogen Treatments in Amorphous In–Ga–Zn–O Films. *ECS J. Solid State Sci. Technol.* **2017**, *6*, P365–P372. [CrossRef]
3. Kamiya, T.; Hosono, H. Material characteristics and applications of transparent amorphous oxide semiconductors. *NPG Asia Mater.* **2010**, *2*, 15–22. [CrossRef]
4. Xiao, X.; Deng, W.; Chi, S.; Shao, Y.; He, X.; Wang, L.; Zhang, S. Effect of O_2 Flow Rate During Channel Layer Deposition on Negative Gate Bias Stress-Induced Vth Shift of a-IGZO TFTs. *IEEE Trans. Electron Devices* **2013**, *60*, 4159–4164. [CrossRef]
5. Lee, H.-J.; Cho, S.H.; Abe, K.; Lee, M.-J.; Jung, M. Impact of transient currents caused by alternating drain stress in oxide semiconductors. *Sci. Rep.* **2017**, *7*, 9782–9790. [CrossRef] [PubMed]
6. Lee, H.-J.; Abe, K.; Cho, S.H.; Kim, J.-S.; Bang, S.; Lee, M.-J. Drain-Induced Barrier Lowering in Oxide Semiconductor Thin-Film Transistors with Asymmetrical Local Density of States. *IEEE J. Electron Devices Soc.* **2018**, *6*, 830–834. [CrossRef]
7. Nomura, K.; Kamiya, T.; Hosono, H. Effects of Diffusion of Hydrogen and Oxygen on Electrical Properties of Amorphous Oxide Semiconductor, In-Ga-Zn-O. *ECS J. Solid State Sci. Technol.* **2013**, *2*, P5–P8. [CrossRef]
8. Bang, J.; Chang, K.J. Diffusion and thermal stability of hydrogen in ZnO. *Appl. Phys. Lett.* **2008**, *92*, 132109. [CrossRef]
9. Van de Walle, C.G. Hydrogen as a Cause of Doping in Zinc Oxide. *Phys. Rev. Lett.* **2000**, *85*, 1012–1015. [CrossRef] [PubMed]
10. Lavrov, E.V.; Herklotz, F.; Weber, J. Identification of two hydrogen donors in ZnO. *Phys. Rev. B* **2009**, *79*, 165210. [CrossRef]
11. Lavrov, E.V.; Herklotz, F.; Weber, J. Identification of Hydrogen Molecules in ZnO. *Phys. Rev. Lett.* **2009**, *102*, 185502. [CrossRef] [PubMed]
12. Nam, Y.; Kim, H.-O.; Cho, S.H.; Park, S.-H. Effect of hydrogen diffusion in an In–Ga–Zn–O thin film transistor with an aluminum oxide gate insulator on its electrical properties. *RSC Adv.* **2018**, *8*, 5622–5628. [CrossRef]
13. Kim, J.; Bang, S.; Lee, S.; Shin, S. A study on H_2 plasma treatment effect on a-IGZO thin film transistor. *J. Mater. Res.* **2012**, *27*, 2318–2325. [CrossRef]
14. Abliz, A.; Gao, Q.; Wan, D.; Liu, X.; Xu, L.; Liu, C.; Li, X.; Chen, H.; Guo, T.; Li, J.; et al. Effects of Nitrogen and Hydrogen Codoping on the Electrical Performance and Reliability of InGaZnO Thin-Film Transistors. *ACS Appl. Mater. Interfaces* **2017**, *9*, 10798–10804. [CrossRef] [PubMed]
15. Tari, A.; Wong, W.S. Effect of dual-dielectric hydrogen-diffusion barrier layers on the performance of low-temperature processed transparent InGaZnO thin-film transistors. *Appl. Phys. Lett.* **2018**, *112*, 073506. [CrossRef]
16. Oh, S.I.; Choi, G.; Hwang, H.; Lu, W.; Jang, J.-H. Hydrogenated IGZO Thin-Film Transistors Using High-Pressure Hydrogen Annealing. *IEEE Trans. Electron Devices* **2013**, *60*, 2537–2541. [CrossRef]
17. Nguyen, T.T.T.; Aventurier, B.; Terlier, T.; Barnes, J.; Templier, F. Impact of Passivation Conditions on Characteristics of Bottom-Gate IGZO Thin-Film Transistors. *J. Disp. Technol.* **2015**, *11*, 554–558. [CrossRef]
18. Han, K.-L.; Ok, K.-C.; Cho, H.-S.; Oh, S.; Park, J.-S. Effect of hydrogen on the device performance and stability characteristics of amorphous InGaZnO thin-film transistors with a $SiO_2/SiN_x/SiO_2$ buffer. *Appl. Phys. Lett.* **2017**, *111*, 063502. [CrossRef]
19. *ATLAS Device Simulation Software User's Manual*; Silvaco International: Santa Clara, CA, USA, 2015.
20. Lee, H.-J.; Abe, K.; Kim, J.S.; Lee, M.-J. Electron-blocking by the potential barrier originated from the asymmetrical local density of state in the oxide semiconductor. *Sci. Rep.* **2017**, *7*, 17963–17970. [CrossRef] [PubMed]
21. Janotti, A.; Van de Walle, C.G. Hydrogen multicentre bonds. *Nat. Mater.* **2007**, *6*, 44–47. [CrossRef] [PubMed]

© 2019 by the authors. Licensee MDPI, Basel, Switzerland. This article is an open access article distributed under the terms and conditions of the Creative Commons Attribution (CC BY) license (http://creativecommons.org/licenses/by/4.0/).

Article

Critical Evaluation of Organic Thin-Film Transistor Models

Markus Krammer [1], James W. Borchert [2], Andreas Petritz [3], Esther Karner-Petritz [3], Gerburg Schider [3], Barbara Stadlober [3], Hagen Klauk [2] and Karin Zojer [1,*]

[1] Institute of Solid State Physics, NAWI Graz, Graz University of Technology, Petersgasse 16, 8010 Graz, Austria; markus.krammer@student.tugraz.at
[2] Max Planck Institute for Solid State Research, Heisenbergstr. 1, 70569 Stuttgart, Germany; J.Borchert@fkf.mpg.de (J.W.B.); H.Klauk@fkf.mpg.de (H.K.)
[3] Joanneum Research Materials, Institute for Surface Technologies and Photonics, Franz-Pichler-Straße 30, 8160 Weiz, Austria; Andreas.Petritz@joanneum.at (A.P.); esther.karner-petritz@joanneum.at (E.K.-P.); Gerburg.Schider@joanneum.at (G.S.); barbara.stadlober@joanneum.at (B.S.)
* Correspondence: karin.zojer@tugraz.at; Tel.: +43-316-873-8974

Received: 20 December 2018; Accepted: 2 February 2019; Published: 6 February 2019

Abstract: The thin-film transistor (TFT) is a popular tool for determining the charge-carrier mobility in semiconductors, as the mobility (and other transistor parameters, such as the contact resistances) can be conveniently extracted from its measured current-voltage characteristics. However, the accuracy of the extracted parameters is quite limited, because their values depend on the extraction technique and on the validity of the underlying transistor model. We propose here a new approach for validating to what extent a chosen transistor model is able to predict correctly the transistor operation. In the two-step fitting approach we have developed, we analyze the measured current-voltage characteristics of a series of TFTs with different channel lengths. In the first step, the transistor parameters are extracted from each individual transistor by fitting the output and transfer characteristics to the transistor model. In the second step, we check whether the channel-length dependence of the extracted parameters is consistent with the underlying model. We present results obtained from organic TFTs fabricated in two different laboratories using two different device architectures, three different organic semiconductors and five different materials combinations for the source and drain contacts. For each set of TFTs, our approach reveals that the state-of-the-art transistor models fail to reproduce correctly the channel-length-dependence of the transistor parameters. Our approach suggests that conventional transistor models require improvements in terms of the charge-carrier-density dependence of the mobility and/or in terms of the consideration of uncompensated charges in the carrier-accumulation channel.

Keywords: organic thin-film transistor; transistor model evaluation; channel-length dependence; contact resistances; modeling contact effects; equivalent circuit; charge-carrier-mobility extraction

1. Introduction

The fabrication of organic thin-film transistors (TFTs) has reached a level at which devices with excellent performance, small device-to-device variations, and smooth electrical characteristics with small hysteresis can routinely be provided [1–4]. These technological advances are significantly ahead of our current ability to reliably extract crucial transistor parameters. Such a reliable extraction procedure is desirable to design integrated circuits, to determine materials parameters, or to optimize the TFT fabrication process. The two most prominent of these transistor parameters are the charge-carrier mobility as a materials parameter and the contact resistance as an indicator for the quality of the contact-semiconductor interfaces. To extract these parameters from the measured current-voltage characteristics, the device operation and, hence, the electrical TFT characteristics must be understood in terms of these parameters.

In general, every parameter extraction approach requires a theoretical model for the transistor operation that provides the current-voltage relations on the basis of input parameters that properly account for the regime of operation (applied voltages), the materials properties, and the device geometry. While materials-related transistor parameters comprise, for example, the charge-carrier mobility and the permittivity of the gate dielectric, the most prominent geometry parameters are the channel length, the channel width, and the gate-dielectric thickness. Such theoretical models hold much promise of being able to associate correctly any changes in the current-voltage characteristics to changes in these transistor parameters. Hence, it is particularly desirable to utilize a theoretical transistor model that associates the drain current of the transistors to these parameters, preferably with a closed analytic expression. To obtain reliable and robust relations, it is customary to conceive of specific models for each class of TFTs by accounting, for example, for a particular transport mechanism [5,6] or particular geometry features, such as a small channel length [7]. The potential success of a theoretical model inherently relies on a set of preliminary assumptions that are guided by the device geometry and by the anticipated transport mechanism. For instance, in the presumably most prominent field-effect-transistor model, the gradual channel approximation, it is assumed that all mobile charges are confined to the interface between the semiconductor layer and the gate dielectric. Despite the many efforts to improve the transistor models in order to better comply with the measured electrical characteristics [8,9], the development of refined models is still in its infancy, as there are no reliable tools to validate the consistency between the prediction made by a given theoretical model and the measured current-voltage characteristics of the transistors.

Here, we propose a new approach for evaluating the adequateness of a suggested theoretical transistor model. Our approach is a two-step process that requires a set of transistors with different channel lengths. The two steps combine the benefits and overcome the drawbacks of the two classes of established extraction approaches, namely the "single transistor methods" and the "channel-length-scaling approaches" [10]. "Single transistor methods" seek to extract the parameters of an assumed transistor model from certain operation regimes in the output or/and transfer characteristics of an individual TFT [9–13], whereas in "channel-length-scaling approaches", parameters are extracted from a series of nominally identical transistors that differ only in the channel length, by exploring the dependence of the transistor characteristics on the channel length from the perspective of the assumed model [14–16]. Neither of these two classes of established extraction approaches is able to validate reliably the consistency between the theoretical model and the measured current-voltage characteristics. For the "single transistor methods", the consistency can, at best, be validated within the limited region from which the transistor parameters are extracted, and for the "channel-length-scaling approaches", the deviations of the model predictions from the measured data are often hidden by unavoidable device-to-device variations.

The approach we present here combines fundamental aspects of these two classes of established parameter extraction methods. This combination allows us to go beyond existing extraction methods by enabling a reliable validation of the adequateness of the theoretical model underlying the extraction method. In the first step of our approach, the current-voltage characteristics of individual transistors are analyzed. We fit the entire set of measured data points of all output and transfer characteristics simultaneously to the underlying theoretical model. As pointed out by Deen et al. [17] and Fischer et al. [18], the simultaneous consideration of all available data points guarantees the best possible parameter set describing an individual transistor as a whole and eliminates the aforementioned ambiguity that arises from selecting certain regions of device operation. The extracted parameter set is then used to calculate the output and transfer characteristics $I_D(V_{DS})$ and $I_D(V_{GS})$. By comparing the calculated and the measured current-voltage characteristics, we are able to perform an initial check of the validity of the underlying theoretical model. Furthermore, any deviations between the calculated and the measured characteristics can be analyzed in order to derive strategies for improving the underlying transistor model. If this check is successful and the calculated characteristics are in good agreement with the measured characteristics, we proceed to the second step of our approach in which we compare the individually-extracted fitting parameters of all devices with regard to their channel-length dependencies. The second step relies on the hypothesis that the transistor can be spatially separated into a charge-accumulation channel region and a source and drain contact region. Within this hypothesis, the contact regions are assumed to behave identical for all transistors, irrespective of the channel length. Only the size of the charge-accumulation channel changes corresponding to the channel length. If the underlying model is able to separate correctly the channel region and the contact regions, the channel-length dependencies of all parameters will be captured explicitly in the model. In turn, all related fitting parameters have to be independent of the channel length. Hence, if in a second check, the extracted fitting parameters are found to be independent of the channel length, we can be certain that the device characteristics are properly and consistently described by the underlying model. This second step is of particular importance, because conventional fitting approaches have the drawback that they routinely produce good agreement between the calculated and the measured characteristics even when the underlying model is unreasonable, as long as a sufficiently large number of parameters is considered [19]. Our two-step fitting approach (TSFA) overcomes these drawbacks of conventional extraction methods and conventional fitting approaches and is thus able to validate even complex theoretical models and identify problems within these models.

We have tested the merit of our TSFA and scrutinized existing organic-TFT models using experimental data. For this purpose, we have selected five sets of organic TFTs. These sets of TFTs differ in the device architecture, the choice of the organic semiconductor, and the functionalization of the contact-semiconductor interface to realize TFTs in which the contact properties range from nearly ideal (very small contact resistances) to highly non-ideal (large, non-linear contact resistances). In particular, we fabricated a set of bottom-gate, bottom-contact TFTs using dinaphtho[2,3-b:2′,3′-f]thieno[3,2-b]thiophene (DNTT) as the semiconductor and Au contacts functionalized with pentafluorobenzenethiol (PFBT) to obtain a small contact resistance [20], a set of bottom-gate, top-contact DNTT TFTs with Au contacts [21], a set of bottom-gate, bottom-contact pentacene TFTs with Au contacts functionalized with 2-phenylpyrimidine-5-thiol (BP0-down) [22], and two sets of bottom-gate, bottom-contact C_{60} TFTs with Au contacts functionalized with either 4-(2-mercaptophenyl)pyrimidine (BP0-up) or biphenyl-4-thiol (BP0) [22]. The DNTT TFTs and the pentacene TFTs are p-channel transistors, while the C_{60} TFTs are n-channel transistors. Since the bottom-gate, bottom-contact DNTT TFTs show almost ideal transistor behavior with very small contact resistances, we have used them as a reference and analyzed them in detail.

We will first explain the application and interpretation of the most popular parameter-extraction method for organic TFTs, the transmission line method (TLM) [14,15]. Second, we will illustrate our TSFA on the example of the theoretical transistor model that underlies the TLM. Third, we will test a more sophisticated transistor model that includes a field- and charge-carrier-density-dependent mobility. These investigations will exclusively use the measured characteristics of the bottom-gate, bottom-contact DNTT TFTs. Finally, we will examine models with field- and charge-carrier-density-dependent mobility and non-linear contact resistances by analyzing the measured current-voltage characteristics of the remaining four sets of TFTs.

2. Materials and Methods

This section discusses (i) how to calculate numerically the drain current within the equivalent circuit model, (ii) how to fit the calculated drain current to the measured data, and (iii) which organic-TFT technologies we have investigated with our TSFA.

2.1. Equivalent Circuit Model

The equivalent circuit model employed in this work is shown in Figure 1a. This model consists of an ideal field-effect transistor in the gradual channel approximation [23] that is characterized by a charge-carrier mobility that depends on the charge-carrier density [24,25] and on the electric field [7,26] and is terminated by the ideal source **S'**, drain **D'**, and gate **G'** terminals. At the gate terminal, the threshold voltage V_T is implemented as an external bias, and the source and drain terminals are connected to Ohmic contact resistances $R_{S,0}$ and $R_{D,0}$. The experimentally-accessible terminals are labeled source **S**, drain **D**, and gate **G**. The assignment of the elements in the equivalent circuit model to the location in a real device is indicated in gray. Figure 1b shows a schematic drawing of a bottom-gate, bottom-contact TFT; Figure 1c shows optical microscopy images of bottom-gate, bottom-contact DNTT TFTs with channel lengths of 2, 8, 40, and 80 µm from top to bottom; and Figure 1d shows a photograph of a set of pentacene TFTs on a flexible plastic substrate.

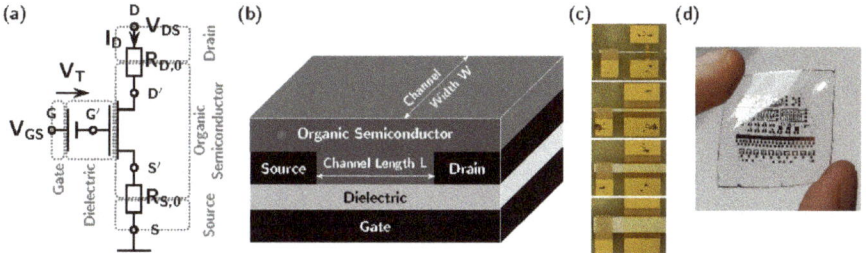

Figure 1. Panel (**a**) shows the equivalent circuit model based on an ideal field-effect transistor in the gradual channel approximation with a field- and charge-carrier-density-dependent mobility connected to the Ohmic source and drain resistances $R_{S,0}$ and $R_{D,0}$. The threshold voltage V_T is implemented in the form of an external bias. The terminals of the ideal transistor are labeled source **S'**, drain **D'**, and gate **G'**, and the experimentally-accessible terminals are labeled source **S**, drain **D**, and gate **G**. The corresponding location of the elements in a real device is indicated in gray. In (**b**), a schematic drawing of a bottom-gate, bottom-contact thin-film transistor (TFT) is illustrated. In (**c**), optical microscopy images of bottom-gate, bottom-contact dinaphtho[2,3-b:2',3'-f]thieno[3,2-b]thiophene (DNTT) TFTs with channel lengths of 2, 8, 40, and 80 µm from top to bottom can be seen, and (**d**) shows a photograph of a set of pentacene TFTs on a flexible plastic substrate.

The charge-carrier mobility μ at a certain position x in the carrier-accumulation channel is determined by:

$$\mu(x) = \mu_0 \exp\left(\beta\sqrt{\frac{L_0}{L}\left|\frac{V_{D'S'}}{V_0}\right|}\right) \left(\frac{V_{GS'} - V_T - V_{ChS'}(x)}{V_0}\right)^\gamma \tag{1}$$

where $V_{ChS'}(x)$ is the channel potential (with respect to the source) at this position x, $V_{G'S'} = V_{GS'} - V_T$ is the gate-source voltage, $V_{D'S'}$ is the drain-source voltage, μ_0 is the mobility prefactor, L is the channel length, β is the exponent of the field sensitivity, γ is the charge-carrier-density sensitivity, $L_0 = 1$ µm is a constant length scale, and V_0 is a constant potential scaling factor, with $V_0 = 1$ V for n-channel (electron-conducting) TFTs and $V_0 = -1$ V for p-channel (hole-conducting) TFTs. Note that the absolute values of the constant length scale L_0 and the constant potential scale V_0 are chosen arbitrarily and are required only to avoid inconsistencies regarding the units within the corresponding power functions. The exponential term mimics a simplified Poole–Frenkel field dependence [7,26], and the right term describes the charge-carrier-density dependence of the mobility with a power law behavior [24,25].

Incorporating the gradual channel approximation (for details, see [8,23]) leads to an implicit system of equations determining the drain current I_D for given applied gate-source voltages and drain-source voltages V_{GS} and V_{DS}:

$$v_{G'S'} = \frac{1}{V_0}\left(V_{GS} - V_T - I_D\frac{r_{S,0}}{W}\right)$$

$$v_{G'D'} = \frac{1}{V_0}\left(V_{GS} - V_T - V_{DS} + I_D\frac{r_{D,0}}{W}\right)$$

$$I_D = \frac{V_0|V_0|WC_I\mu_0}{L(\gamma+2)}\exp\left(\beta\sqrt{\frac{L_0}{L}}|v_{G'S'} - v_{G'D'}|\right)\left[v_{G'S'}^{\gamma+2}\Theta(v_{G'S'}) - v_{G'D'}^{\gamma+2}\Theta(v_{G'D'})\right] \quad (2)$$

The reduced voltages $v_{G'S'}$ and $v_{G'D'}$ are the voltages between the ideal gate **G'**, source **S'**, and drain **D'** terminals divided by V_0. The Heaviside function $\Theta(x)$ is equal to 1 for $x \geq 0$ and equal to 0 for $x < 0$. C_I is the gate capacitance per unit area, and $r_{S,0} = R_{S,0}W$ and $r_{D,0} = R_{D,0}W$ are the channel-width-normalized source and drain resistances, respectively. The drain current I_D as the output parameter is thus implicitly determined by two input parameters V_{GS} and V_{DS}, six fitting parameters V_T, μ_0, $r_{S,0}$, $r_{D,0}$, β, and γ, two constants L_0 and V_0, and three geometry parameters L, W, and C_I. The gate capacitance per unit area C_I is considered here as a geometry parameter since it is determined by the thickness and the permittivity of the gate dielectric.

The implicit system of Equations (2) can be numerically solved with the bisection method, incorporating knowledge of the desired fixed point. We start by setting $I_D^{(0)} = 0$ A in the first two equations of (2) to obtain $v_{G'S'}^{(0)}$ and $v_{G'D'}^{(0)}$ and then substituting the latter in the right-hand side of the third equation. This yields $I_D^{(1)}$ and defines the search interval $[I_{D,min}, I_{D,max}] = [\min(I_D^{(0)}, I_D^{(1)}), \max(I_D^{(0)}, I_D^{(1)})]$. Now, the recurrent series starts by taking the midpoint $I_{D,MP} = (I_{D,min} + I_{D,max})/2$ and plugging it into the first two equations and the right-hand side of the third equation of (2) to get $I_{D,calc}$. If $I_{D,MP} < I_{D,calc}$, the new search interval is $[I_{D,MP}, \min(I_{D,max}, I_{D,calc})]$, and if $I_{D,MP} > I_{D,calc}$, the new search interval is $[\max(I_{D,min}, I_{D,calc}), I_{D,MP}]$. Calculating $I_{D,MP}$ and $I_{D,calc}$ is continued until the desired accuracy is reached.

2.2. Fitting Procedure

The fitting of the measured current-voltage characteristics to this model is accomplished by applying a Gauss–Newton algorithm with the variation of Marquardt [27]. The algorithm is slightly modified here so that it is able to handle minimum and maximum parameter values. In our case, μ_0, $r_{S,0}$, $r_{D,0}$, and β have to be positive and γ must be greater than -1. The Gauss–Newton–Marquardt algorithm calculates the difference $\Delta a = a - a^{(0)}$ between the previous model parameters $a^{(0)}$ and the suggested new model parameters a by solving the system of linear equations:

$$(A + \lambda D)\Delta a = b \quad (3)$$

with matrices A and D, the convergence parameter λ introduced by Marquardt, and a vector b. The matrix A is given by:

$$(A)_{ij} = \sum_{k=1}^{n}\frac{1}{\sigma_k^2}\frac{\partial I_D(V_{DS}^{(k)}, V_{GS}^{(k)}; a^{(0)})}{\partial a_i}\frac{\partial I_D(V_{DS}^{(k)}, V_{GS}^{(k)}; a^{(0)})}{\partial a_j}, \quad (4)$$

containing the sum over all n measured values k, the standard deviation σ_k, and the partial derivatives $\partial I_D(V_{DS}^{(k)}, V_{GS}^{(k)}; a^{(0)})/\partial a_{i/j}$ of the calculated drain current I_D at the measured data values $V_{DS}^{(k)}$ and $V_{GS}^{(k)}$ and the previous model parameters $a^{(0)}$ with respect to the model parameters a_i and a_j, respectively.

The matrix D is a diagonal matrix consisting of the diagonal elements of A, $(D)_{ij} = \delta_{ij}(A)_{ij}$ with δ_{ij} being the Kronecker delta returning 1 if $i = j$ and 0 if $i \neq j$. The vector b is given by:

$$b_i = \sum_{k=1}^{n} \frac{I_D^{(k)} - I_D(V_{DS}^{(k)}, V_{GS}^{(k)}; a^{(0)})}{\sigma_k^2} \frac{\partial I_D(V_{DS}^{(k)}, V_{GS}^{(k)}; a^{(0)})}{\partial a_i} \tag{5}$$

where $I_D^{(k)}$ is the drain current measured at the drain-source and gate-source voltages $V_{DS}^{(k)}$ and $V_{GS}^{(k)}$.

To consider minimum and maximum values of the model parameters, the matrices A and D, the vector b, and the convergence parameter λ are evaluated as in [27], and the system of linear Equation (3) is solved to determine Δa. Before continuing with this calculated value for Δa, we need to check whether any of the suggested parameters $a = a^{(0)} + \Delta a$ are out of bounds. For all entries j that are out of bounds, Δa_j is changed so that a_j stays within bounds (e.g., $\Delta a_j = a_j^{max} - a_j^{(0)}$ if the upper boundary is exceeded) and substituted in the system of linear Equation (3) by eliminating the corresponding equation j and transferring $(A)_{ij}\Delta a_j$ to the right-hand side $b_i \rightarrow b_i - (A)_{ij}\Delta a_j$. The new system of linear equations is solved, and the model parameters are checked again. This procedure is iteratively continued until all model parameters are within bounds. At this point, the Gauss–Newton algorithm is continued.

To calculate the required derivatives of the model function with respect to the model parameters, a few definitions are useful:

$$T_0 = \beta\sqrt{\frac{L_0}{L}} \frac{v_{G'S'}^{\gamma+2}\Theta(v_{G'S'}) - v_{G'D'}^{\gamma+2}\Theta(v_{G'D'})}{2(\gamma+2)\sqrt{|v_{G'S'} - v_{G'D'}|}} \operatorname{sgn}(v_{G'S'} - v_{G'D'}), \tag{6}$$

$$T_{G'S'} = v_{G'S'}^{\gamma+1}\Theta(v_{G'S'}) + T_0, \tag{7}$$

$$T_{G'D'} = v_{G'D'}^{\gamma+1}\Theta(v_{G'D'}) + T_0, \tag{8}$$

$$\tilde{\mu}_0 = \mu_0 \exp\left(\beta\sqrt{\frac{L_0}{L}}|v_{G'S'} - v_{G'D'}|\right), \tag{9}$$

$$D_{I_D} = 1 + \frac{|V_0|C_I\tilde{\mu}_0}{L}(T_{G'S'}r_{S,0} + T_{G'D'}r_{D,0}) \tag{10}$$

The sign function $\operatorname{sgn}(x)$ is equal to -1 if $x < 0$, equal to 1 if $x > 0$, and equal 0 if $x = 0$. With these definitions, the derivatives can be written in a compact way as follows:

$$\frac{\partial I_D}{\partial V_T} = -\frac{|V_0|WC_I\tilde{\mu}_0}{LD_{I_D}}(T_{G'S'} - T_{G'D'}) \tag{11}$$

$$\frac{\partial I_D}{\partial \mu_0} = \frac{I_D}{\mu_0 D_{I_D}} \tag{12}$$

$$\frac{\partial I_D}{\partial r_{S,0}} = -\frac{|V_0|C_I\tilde{\mu}_0 T_{G'S'} I_D}{LD_{I_D}} \tag{13}$$

$$\frac{\partial I_D}{\partial r_{D,0}} = -\frac{|V_0|C_I\tilde{\mu}_0 T_{G'D'} I_D}{LD_{I_D}} \tag{14}$$

$$\frac{\partial I_D}{\partial \gamma} = -\frac{I_D}{D_{I_D}(\gamma+2)} - \frac{V_0|V_0|WC_I\tilde{\mu}_0}{L(\gamma+2)D_{I_D}}\left[\ln(v_{G'S'})v_{G'S'}^{\gamma+2}\Theta(v_{G'S'}) - \ln(v_{G'D'})v_{G'D'}^{\gamma+2}\Theta(v_{G'D'})\right] \tag{15}$$

$$\frac{\partial I_D}{\partial \beta} = \frac{I_D}{D_{I_D}}\sqrt{\frac{L_0}{L}}|v_{G'S'} - v_{G'D'}| \tag{16}$$

In addition to these derivatives, the start values for the fitting procedure are required. Initially, we can set all parameters to zero, except for the mobility prefactor μ_0 and the threshold voltage V_T.

These two parameters can be estimated from the saturation regime of the output characteristics. In this regime and with only μ_0 and V_T being non-zero, the drain current I_D is calculated as $I_{D,sat} = WC_I\mu_0(V_{GS} - V_T)^2/2L$. Performing a linear fit of $\sqrt{I_{D,sat}}(V_{GS})$ provides start values for μ_0 and V_T. With these start values, the initial fitting is performed by optimizing only μ_0 and V_T. From these optimized parameters, more and more parameters are included in the fitting procedure. The next fit, e.g., is to optimize μ_0, V_T, $r_{S,0}$, and $r_{D,0}$ followed by a fit of μ_0, V_T, $r_{S,0}$, $r_{D,0}$, and γ, and a final fit of μ_0, V_T, $r_{S,0}$, $r_{D,0}$, γ, and β. When changing the order of the parameters included in the fitting procedure (e.g., β before γ), the optimized parameters should converge to the same solution within the chosen numerical accuracy.

2.3. Device Fabrication

All TFTs were fabricated on flexible plastic substrates using aluminum oxide as the gate dielectric. Details regarding the device architecture and the materials employed for the semiconductor and the source and drain contacts, the gate-dielectric thickness and the channel lengths and channel widths can be found in Table 1. The TFTs investigated in detail are bottom-gate, bottom-contact TFTs with a 30 nm-thick layer of DNTT as the semiconductor and Au source and drain contacts functionalized with PFBT to increase the work function of the contacts [28] and to optimize the semiconductor morphology across the contact interface [20]. The 5.3 nm-thick aluminum oxide gate dielectric enables operation voltages below 3 V [29]. This set of TFTs was chosen because the current-voltage characteristics of these TFTs most closely resemble those of an ideal field-effect transistor, as indicated by a nearly perfectly linear relation between the measured drain current and the applied drain-source voltage at small drain-source voltages (i.e., in the linear regime of operation), small contact resistances, and very good device-to-device uniformity. This nearly ideal current-voltage behavior is present even in the TFTs with the smallest channel length implemented here ($L = 2$ μm).

The TFTs of the remaining four sets of devices (i.e., the bottom-gate, top-contact DNTT TFTs [21], the bottom-gate, bottom-contact pentacene TFTs [22], and the bottom-gate, bottom-contact C_{60} TFTs [22]) will also be analyzed, albeit only briefly.

Table 1. Device architecture, materials employed for the organic semiconductor and the source and drain contacts, gate-dielectric thickness $d_{Al_2O_3}$, and the range of channel lengths L and channel widths W of the TFTs analyzed in this work. Device architectures are the bottom-gate, bottom-contact (BGBC) and the bottom-gate, top-contact (BGTC) structure. The Au contacts of the BGBC TFTs were functionalized with either pentafluorobenzenethiol (PFBT), 2-phenylpyrimidine-5-thiol (BP0-down), 4-(2-mercaptophenyl)pyrimidine (BP0-up), or biphenyl-4-thiol (BP0). DNTT TFTs with channel lengths $L \leq 4$ μm have a channel width of 20 μm, and DNTT TFTs with channel lengths $L > 4$ μm have a channel width of 200 μm.

Name/Reference	Architecture	Semiconductor	Contact	$d_{Al_2O_3}$ (nm)	L (μm)	W (μm)
DNTT-BC [29]	BGBC	DNTT	Au/PFBT	5.3	2–80	20–200
DNTT-TC [21]	BGTC	DNTT	Au	5.3	4–100	20–200
Pentacene [22]	BGBC	Pentacene	Au/BP0-down	18	4.85–52.90	1000
C_{60}-BP0-up [22]	BGBC	C_{60}	Au/BP0-up	18	3.0–100.5	1000
C_{60}-BP0 [22]	BGBC	C_{60}	Au/BP0	18	3.6–51.0	1000

3. Results

3.1. Conventional Transmission Line Method

Before applying our two-step fitting approach (TSFA), we analyze the data measured on the bottom-contact DNTT TFTs using the popular transmission line method (TLM). This analysis is performed to (i) put the measured current-voltage characteristics into a perspective commonly shared in our field of research and (ii) highlight the benefits and drawbacks of the conventional TLM.

In principle, the TLM can also handle certain non-idealities, such as non-Ohmic contact resistances. However, when applying the most commonly-used extraction procedure proposed in the conventional TLM, the model assumptions are rather strict and require (i) an ideal field-effect transistor in the gradual channel approximation [23] having a charge-carrier mobility that is independent of the electric fields and the charge-carrier density and (ii) Ohmic source and drain resistances [14,15]. Under these model assumptions, the drain current I_D in the linear regime of the output characteristics is implicitly determined by:

$$I_D = \frac{V_0 W C_I \mu_{TLM}}{2|V_0|(L+L_T)} \left[\left(V_{GS} - V_T - I_D \frac{r_{S,0}}{W} \right)^2 - \left(V_{GS} - V_T - V_{DS} + I_D \frac{r_{D,0}}{W} \right)^2 \right]. \tag{17}$$

The transfer length L_T accounts for a channel-length-independent extension of the charge-accumulation channel at the contacts. In bottom-gate, top-contact TFTs, L_T can be interpreted as an additional distance that the charge carriers travel laterally through the organic semiconductor layer underneath the contacts to reach the charge-accumulation channel (or the drain contact) (see, e.g., [18]). In bottom-gate, bottom-contact TFTs, charges are injected and extracted very close to the channel, so the distances that the carriers travel above the contacts before reaching the channel (or the drain) are very small. This implies that in bottom-gate, bottom-contact TFTs, L_T is not necessarily a physically-interpretable parameter, but rather has to be seen as a weighting factor for a non-Ohmic contribution to the contact resistance.

The parameter extraction procedure consists of three parts. In the first part, the on-state resistance r_{on} is calculated from the slope of the measured output characteristics:

$$r_{on} = \lim_{V_{DS} \to 0} W \frac{\partial V_{DS}}{\partial I_D} = |V_0| \frac{L+L_T}{V_0 C_I \mu_{TLM}(V_{GS}-V_T)} + r_{C,0} \tag{18}$$

with $r_{C,0} = r_{S,0} + r_{D,0}$. Note that it is important to extract r_{on} for $V_{DS} \to 0$ V because only at this point is it possible to separate the contacts clearly from the channel within the model (cf. Figure S1 in the Supplementary Materials). To determine r_{on}, we performed a linear fit to the output curves measured for the four smallest drain-source voltages and forced this fit to pass through the origin at $V_{DS} = 0$ V and $I_D = 0$ A. The plot of the on-state resistance r_{on} as a function of the channel length L for different gate-source voltages is shown in Figure 2a. As can be seen, the measured on-state resistance is indeed proportional to the channel length, and the linear fits to the measured on-state resistance at different gate-source voltages intersect approximately at $L \approx -3.2$ μm and $r_{on} \approx 0.15$ kΩcm.

In the second part of the TLM analysis, the inverse slope $\Delta L/\Delta r_{on} = C_I \mu_{TLM}(V_{GS}-V_T)V_0/|V_0|$ is extracted from Figure 2a and plotted as a function of the gate-source voltage V_{GS} (see Figure 2b). The slope of this plot yields the intrinsic channel mobility $\mu_{TLM} = 3.2$ cm²/Vs, and the x-axis intersect yields the threshold voltage $V_T = -1.25$ V. In the last part of the TLM analysis, the on-state resistance at a channel length of zero, $r_{on}(L=0) = r_{Sh}L_T + r_{C,0}$, is plotted as a function of the sheet resistance $r_{Sh} = |V_0| [V_0 C_I \mu_{TLM}(V_{GS}-V_T)]^{-1}$ (see Figure 2c). The slope of the linear fit to this data is the transfer length $L_T = 3.4$ μm, and the y-axis intersect yields the Ohmic contact resistance $r_{C,0} = 0.14$ kΩcm.

The parameters extracted from the conventional TLM analysis can be considered to be reliable only if the following requirements are fulfilled:

- The measured output characteristics (gray symbols in Figure 3a–d) must be linear for very small drain-source voltages ($V_{DS} \to 0$ V), and the slope of the curves must decrease monotonically as the absolute value of the drain-source voltage V_{DS} increases. An S-shape of the output curves in this regime is an indicator of a non-Ohmic contact resistance.
- The relations r_{on} versus L (Figure 2a), $\Delta L/\Delta r_{on}$ versus V_{GS} (Figure 2b), and $r_{on}(L=0)$ versus r_{Sh} (Figure 2c) must be linear.
- The values for the transfer length L_T and the total Ohmic contact resistance $r_{C,0}$ obtained from the plot $r_{on}(L=0)$ versus r_{Sh} (Figure 2c) must be equal to the values for L and r_{on} at the intersection in Figure 2a.

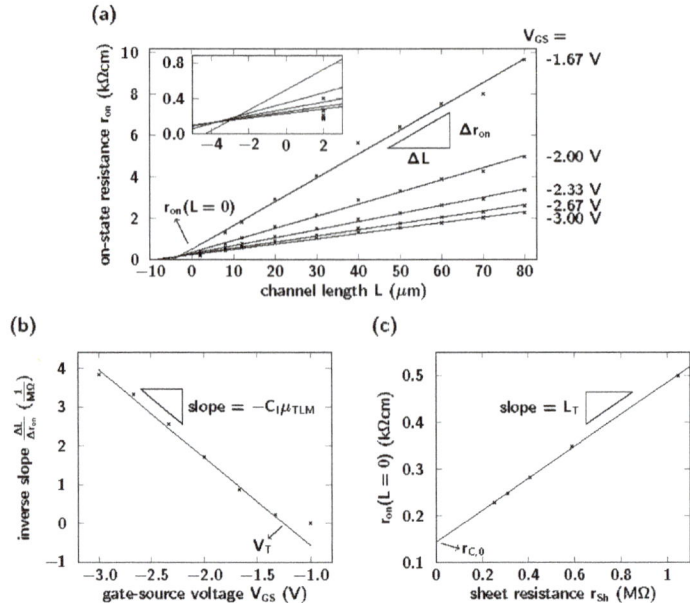

Figure 2. Parameter extraction in the framework of the conventional transmission line method (TLM), performed here on a set of bottom-gate, bottom-contact p-channel TFTs based on the small-molecule semiconductor DNTT. In (**a**), the on-state resistance $r_{on} = W \partial V_{DS}/\partial I_D$ for $V_{DS} \to 0$ V, extracted from the measured output characteristics, is plotted as a function of the channel length for different gate-source voltages V_{GS}. From a linear fit to the data, the inverse slope $\Delta L/\Delta r_{on}$ and the y-axis intersect $r_{on}(L=0)$ are extracted. The inset shows a magnification of the region in which the linear fits intersect and the extracted r_{on} values for the smallest channel length of $L = 2$ μm (symbols). In (**b**), $\Delta L/\Delta r_{on}$ is plotted as a function of the gate-source voltage V_{GS}, and from the linear fit to the data, the threshold voltage $V_T = 1.25$ V and the intrinsic channel mobility $\mu_{TLM} = 3.2$ cm^2/Vs are obtained. In (**c**), $r_{on}(L=0) = r_{Sh}L_T + r_{C,0}$ is plotted as a function of the sheet resistance $r_{Sh} = |V_0| \left[V_0 C_I \mu_{TLM}(V_{GS} - V_T) \right]^{-1}$, and from the linear fit to the data, the transfer length $L_T = 3.4$ μm and the total Ohmic contact resistance $r_{C,0} = 0.14$ kΩcm are obtained.

Figures 2 and 3 confirm that all of these requirements are indeed fulfilled for our set of bottom-gate, bottom-contact DNTT TFTs. Small deviations of the on-state resistances r_{on} extracted for different channel lengths from the linear fit (see Figure 2a) can be attributed to device-to-device variations. A closer look, however, reveals other, more serious inconsistencies. The inset in Figure 2a shows a close-up of the r_{on} versus L relation close to $L = 0$ together with the on-state resistances r_{on} for the smallest channel length $L = 2$ μm (symbols). As can be seen, the linear fits to the r_{on} versus L data do not intersect in a single point. In addition, all the on-state resistances r_{on} extracted from the data of the TFT with the smallest channel length ($L = 2$ μm) are a factor of approximately two below the corresponding linear fits. These two inconsistencies do not invalidate a further analysis, because the fact that the linear fits to the r_{on} versus L data do not intersect in a single point could just be a consequence of the drain-source voltage V_{DS} being too large to be able to extract the on-state resistance r_{on} in a reliable manner (cf. Figure S1), and the deviation of the on-state resistances r_{on} extracted from the data of the TFT with the channel length of $L = 2$ μm might be caused by short-channel effects. These explanations do not necessarily compromise the validity of the model system.

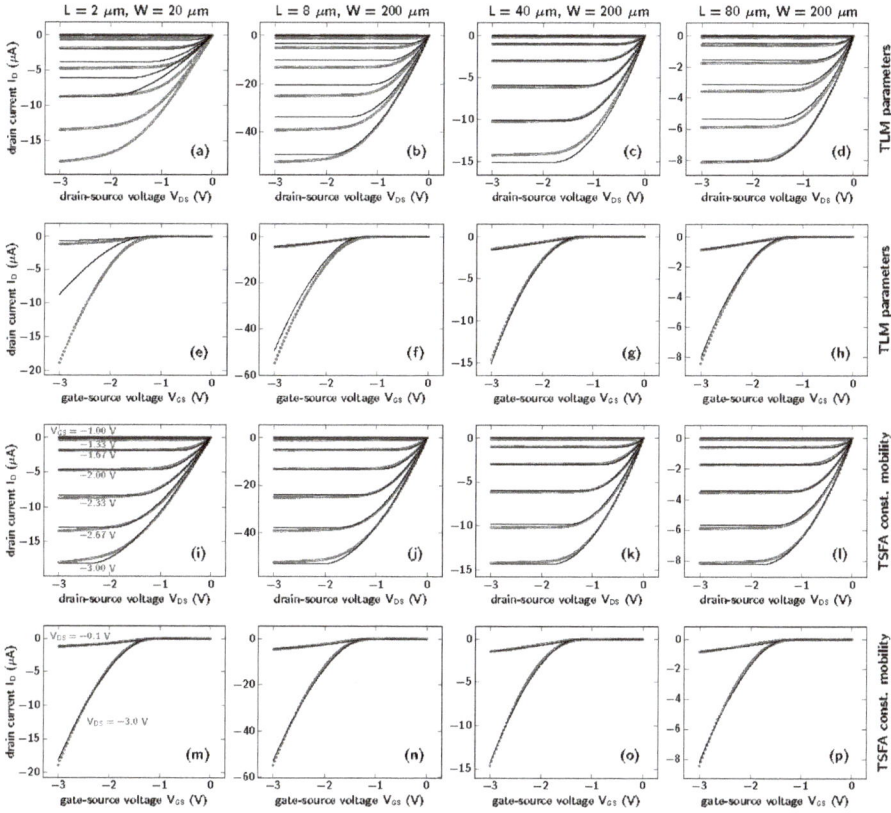

Figure 3. Measured output and transfer characteristics (gray symbols) and calculated output and transfer characteristics (black lines) of bottom-gate, bottom-contact DNTT TFTs with channel lengths L of 2, 8, 40, and 80 µm. The TFT with a channel length of 2 µm has a channel width of 20 µm, while the TFTs with channel lengths of 8, 40, and 80 µm have a channel width of 200 µm. Note that the gray symbols appear as an apparent thick line due to the close spacing of the data points. In (**a–h**), the output and transfer curves were calculated using the transistor parameters extracted using the conventional TLM analysis. In (**i–p**), the output and transfer curves were calculated using our two-step fitting approach (TSFA) with the constant-mobility model underlying the conventional TLM.

As we are able to calculate the electrical TFT characteristics for a given set of device parameters, we can now compare the output and transfer characteristics calculated using the parameters extracted from the TLM analysis to the measured output and transfer characteristics. This comparison is shown in Figure 3a–h for TFTs with channel lengths of 2, 8, 40, and 80 µm. As can be seen, the calculated output curves (black lines) deviate substantially from the measured output curves (gray symbols), regardless of the channel length. These deviations indicate a problem within the transistor model underlying the TLM that had evaded the reliability check performed above. Upon closer inspection, it can be noticed that the agreement between the calculated and measured output and transfer curves is particularly poor when the channel length is small (Figure 3a,e). For the three larger channel lengths (8, 40, and 80 µm; Figure 3b–d), the slope of the output curves at small drain-source voltages (linear regime) is captured reasonably well, but the agreement becomes increasingly worse with increasing absolute value of the drain-source voltage (saturation regime). The better agreement between the calculated and the measured output curves at small drain-source voltages (linear regime) is due to the fact that the transistor parameters in the TLM analysis were extracted for $V_{DS} \to 0$ V.

3.2. TSFA with Constant-Mobility Model

The conventional TLM analysis is able to produce reliable results only if all model parameters are identical for all transistors within the set of devices with different channel lengths. This is a substantial weakness of the conventional TLM, because in reality, these parameters can vary considerably, even for nominally identical organic transistors. Such device-to-device variations may explain the deviations between the calculated and the measured output and transfer characteristics, as seen in Figure 3a–h. Therefore, the question arises whether these deviations can be attributed to the extraction method (TLM) or to the underlying transistor model. To answer this question, we have analyzed the measured TFT data using our TSFA. We have extracted a charge-carrier mobility μ_{TSFA}, a threshold voltage V_T, and source and drain resistances $r_{S,0}$ and $r_{D,0}$ for each TFT individually. Note that the transfer length L_T cannot be evaluated in the first step of the TSFA, and the charge-carrier mobility extracted within the TSFA is related to the intrinsic channel mobility μ_{TLM} extracted using the TLM as $\mu_{TSFA}/L = \mu_{TLM}/(L + L_T)$. The output and transfer characteristics calculated with this approach are shown in Figure 3i–p. For all channel lengths, the agreement between the output and transfer curves calculated using our TSFA and the measured output and transfer curves (Figure 3i–p) is significantly better than the agreement between the output and transfer curves calculated using the parameters obtained from the conventional TLM analysis and the measured output and transfer curves (Figure 3a–h). The deviations between the calculated and the measured output characteristics seen in Figure 3i–l are discussed in detail below. Since these deviations are relatively small and the deviations in the transfer curves (Figure 3m–p) are even smaller, the main message taken from Figure 3i–p is that the first step of our TSFA is conditionally passed.

For the second step of our TSFA, we plot the extracted transistor parameters as a function of the channel length, as shown in Figure 4. To be consistent with the model assumptions, these parameters would have to be independent of the channel length L. Figure 4 shows that this is clearly not the case. In Figure 4a, it can be seen that the absolute value of the threshold voltage V_T decreases by about 100 mV as the channel length is decreased from 40 μm–2 μm. This is the well-known threshold-voltage roll-off that occurs in all field-effect transistors (cf. [30], Chapter 6.4.2). Figure 4b shows a pronounced dependence of the charge-carrier mobility μ_{TSFA} on the channel length L (symbols). If we were to strictly stick to the model underlying the TLM, we could surmise that this dependence might be related to the transfer length L_T. To check whether the introduction of a transfer length conceptually lifts the observed channel-length dependence, we can incorporate L_T into the second step of the TSFA by replacing the mobility μ_{TSFA} with the term $\mu_{TLM}\frac{L}{L+L_T}$, where μ_{TLM} should be independent of the channel length. This relation is reminiscent of, but not equivalent to the relation between effective mobility and intrinsic channel mobility (cf. [21,31]). A fit to the relation $\mu_{TSFA} = \mu_{TLM}\frac{L}{L+L_T}$ is shown as a solid line in Figure 4b. As can be seen, the agreement between the fit and the data is quite poor, as the fit systematically overestimates the extracted parameters for intermediate channel lengths ($8 \leq L \leq 40$ μm) and underestimates them for large channel lengths ($L \geq 50$ μm). This poor agreement indicates a problem with the model system. Figure 4c displays the combined contact resistance $r_{C,0} = r_{S,0} + r_{D,0}$. Rather than being independent of the channel length, the contact resistance $r_{C,0}$ increases by more than a factor of three with increasing channel length, which is a clear indicator of an inadequate transistor model.

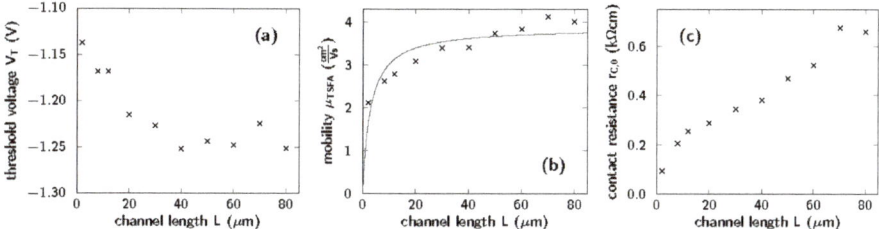

Figure 4. Dependence of the transistor parameters extracted using the TSFA for the theoretical transistor model underlying the conventional TLM on the channel length L. In (**a**), it can be seen that the threshold voltage V_T shows only a very small dependence on the channel length. Panel (**b**) indicates that the dependence of the charge-carrier mobility μ_{TSFA} on the channel length L (symbols) is not properly described by the equation $\mu_{TSFA} = \mu_{TLM} L/(L + L_T)$ (solid line), with L_T being the transfer length. In (**c**), the distinct linear increase of the contact resistance $r_{C,0} = r_{S,0} + r_{D,0}$ with increasing channel length L cannot be explained at all. As a consequence, the model does not pass the second step.

An explanation for the failure of the constant-mobility model can be found by taking a closer look at the deviations between the calculated and the measured output characteristics in Figure 3i–l. We notice that neither the transition between the linear regime and the saturation regime, nor the saturation of the drain current at large drain-source voltages in the measured output curves are properly reproduced in the calculated output curves. The first of these symptoms occurs regardless of the channel length and can be alleviated by assuming that the charge-carrier mobility is a function of the charge-carrier density of the form $\mu = \mu_0 (V_G - V_{Ch})^\gamma$, as suggested by the percolation theory [24] or by the multiple trapping and release model [25]. The second symptom is more pronounced for shorter channels, which indicates a field-dependence of the charge-carrier mobility. As a first attempt, we assume a simplified Poole–Frenkel behavior of the form $\exp(\beta\sqrt{V_{DS}/L})$ [7,26].

3.3. TSFA with Field- and Charge-Carrier-Density-Dependent Mobility

Incorporating a field- and charge-carrier-density-dependent mobility in the model leads to a significantly better agreement between the calculated and the measured output and transfer characteristics (Figure 5a–h) compared to the constant-mobility model (cf. Figure 3i–p). Especially for the TFT with the smallest channel length (Figures 3i,m and 5a,e), the agreement is substantially improved due to the fact that the Poole–Frenkel model provides a far more realistic description of the saturation regime. For all channel lengths, the agreement between the calculated and the measured output curves (Figure 5a–d) is nearly perfect for the smaller gate-source voltages ($|V_{GS}| < 2.5$ V). For the transfer curves (Figure 5e–h), a slight improvement at the branching point at a gate-source voltage of about $V_{GS} = -1.5$ V can be seen compared to the constant-mobility model (Figure 3m–p). We again move on to examine the channel-length dependence of the extracted parameters. The most relevant parameters are the mobility prefactor μ_0 and the combined contact resistance $r_{C,0} = r_{S,0} + r_{D,0}$ shown in Figure 5i,j. The mobility prefactor μ_0 exhibits a slightly smaller channel-length dependence compared to the charge-carrier mobility μ_{TSFA} examined earlier (cf. Figure 4b). The channel-length dependence of $r_{C,0}$ is even more pronounced, with a ratio of approximately one order of magnitude between that of the TFT with the largest channel length and that of the TFT with the smallest channel length (see Figure 5j), causing this model to fail. To illustrate the significant influence of the channel-length-dependence of the contact resistance, Figure S2 shows the disagreement between the calculated and the measured output characteristics when considering the contact resistances of the TFTs with the smallest channel length (Figure S2a–d) and of the TFTs with the largest channel length (Figure S2e–h). The remaining parameters, V_T, γ, and β do not have such a pronounced channel-length dependence (not shown).

To identify the problem of the model, we again inspect the calculated and the measured output characteristics (Figure 5a–d). For all channel lengths, the output curves calculated for $V_{GS} = -2.67$ V lie above and the output curves calculated for $V_{GS} = -3.00$ V lie below the measured output curves. This inaccurate spacing of the curves in the saturation regime is an indicator for a problem of the charge-carrier-density dependence of the charge-carrier mobility, which is predominantly determined by the gate-source voltage V_{GS}. The spacing of the curves in the saturation regime is determined not only by the charge-carrier-density dependence of the mobility, but also by the contact resistances (explained in more detail in Figure S3). Assuming a constant mobility and zero contact resistance, the saturation current $I_{D,sat}$ increases quadratically with the gate-source voltage, $(V_{GS} - V_T)^2$. On the other hand, assuming a constant mobility and a very large contact resistance, the saturation current would increase linearly with increasing gate-source voltage. This means that increasing both the mobility and the contact resistance can lead to similar output curves for the largest gate-source voltage and different spacings for smaller gate-source voltages (see Figure S3).

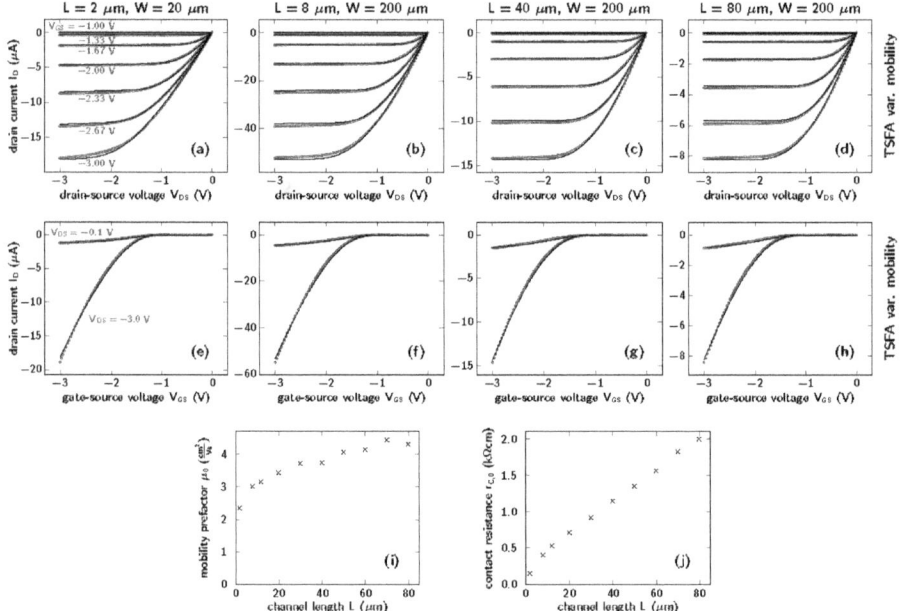

Figure 5. Results of the TSFA for the model with a field- and charge-carrier-density-dependent mobility. In (**a–h**), the measured output and transfer characteristics (gray symbols) and the calculated output and transfer characteristics (black lines) of bottom-gate, bottom-contact DNTT TFTs with channel lengths L of 2, 8, 40, and 80 µm and channel widths W of 20 or 200 µm are shown, indicating good agreement. Note that the gray symbols appear as an apparent thick line due to the close spacing of the data points. In (**i,j**), the channel-length dependence of the mobility prefactor μ_0 and the combined contact resistance $r_{C,0} = r_{S,0} + r_{D,0}$ indicates a failure of the model.

This effect may explain the increase of $r_{C,0}$ with L in the following way: If the charge-carrier-density dependence of the charge-carrier mobility is captured incorrectly, the spacing of the output characteristics for different gate-source voltages will be inaccurate, as well. The incorrect spacing can be compensated by a correspondingly incorrect choice of the contact resistances. As the error in the mobility scales with the channel length in the calculation of the drain current because it is a channel property and the influence of the contact resistance on the drain current is not affected by the channel length, the value extracted for the contact resistance is forced to scale with L in order to compensate the incorrect mobility.

The over- and under-estimation of the drain current for the second-most-negative and the most-negative gate-source voltage suggests that the contact resistance decreases the spacing for more-negative gate-source voltages and, hence, is too large. This change in spacing can alternatively be achieved if the mobility would decrease with increasing charge-carrier density. This decrease should occur only for large charge-carrier densities, because for small charge-carrier densities, i.e., at small gate-source voltages, the increasing mobility in the improved TFT model describes the measured output curves substantially better than the constant-mobility model. Thus, the evaluation of our TSFA suggests that the mobility should first increase and then decrease as the charge-carrier density is increased. Indications for such a behavior of the mobility were recently found experimentally by Bittle et al. [32] and Uemura et al. [33]; Fishchuk et al. [34] suggested such a behavior from a theoretical point of view.

Besides improving the mobility, another possible problem with the transistor model is that the gradual channel approximation does not take into account that organic semiconductors are in principle electrical insulators and that all mobile charges have to be provided by the metal contacts. Unlike in field-effect transistors based on doped semiconductors, such as silicon MOSFETs, these mobile charges in organic semiconductors are not compensated in the semiconductor by charges of opposite polarity. This uncompensated charge accumulation affects the electric field at the source and drain contacts, and this effect becomes more pronounced with increasing channel length. Including this charge cloud in the transistor model might also help alleviate the channel-length dependence of the contact resistance.

3.4. Testing Other Organic-TFT Technologies

We note that the failure of the transistor model discussed above is not exclusive to the bottom-gate, bottom-contact DNTT TFTs investigated above, as the model has also failed for the bottom-gate, top-contact DNTT TFTs [21] and the bottom-gate, bottom-contact pentacene and C_{60} TFTs [22]. For the bottom-gate, top-contact DNTT TFTs and the bottom-gate, bottom-contact C_{60} TFTs, we were able to obtain acceptable agreement between the calculated and the measured current-voltage characteristics by modeling the non-linearity in the linear regime of the output characteristics using a gate-voltage-dependent Schottky diode at the source contact [22]. Selected examples of calculated and measured output characteristics for each set of TFTs are shown in Figure 6a–d. The deviations between the calculated output characteristics and the measured output characteristics are similar to those for the bottom-gate, bottom-contact DNTT TFTs, exhibiting an overestimation of the absolute value of the drain current for the largest absolute value of the gate-source voltage and an underestimation for the smaller ones.

Figure 6e–h shows the Ohmic component of the combined contact resistance $r_{C,0}$ as a function of the channel length L for the bottom-gate, top-contact DNTT TFTs (Figure 6e), the bottom-gate, bottom-contact pentacene TFTs with Au contacts functionalized with 2-phenylpyrimidine-5-thiol (BP0-down) (Figure 6f), and the bottom-gate, bottom-contact C_{60} TFTs with Au contacts functionalized with either 4-(2-mercaptophenyl)pyrimidine (BP0-up) or biphenyl-4-thiol (BP0) (Figure 6g,h). The approximately linear dependence of $r_{C,0}$ on the channel length causes a similar failure of the transistor model during the second step of our TSFA for each set of TFTs.

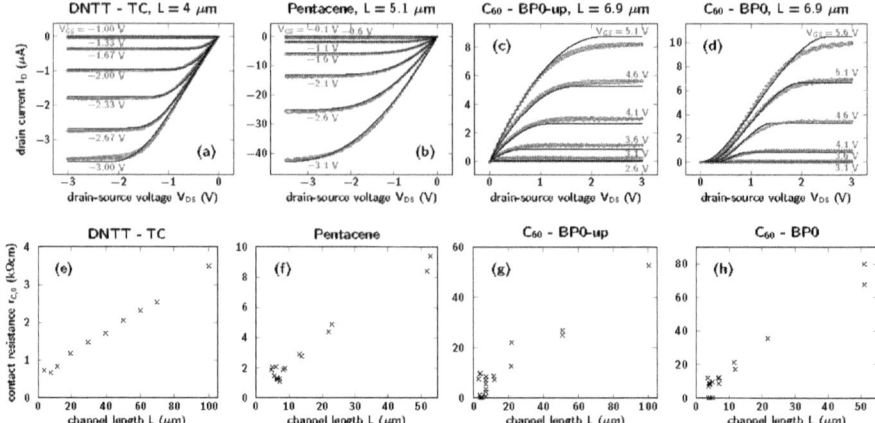

Figure 6. Selected examples of output characteristics for each set of TFTs (**a**–**d**) and Ohmic component of the combined contact resistance $r_{C,0}$ plotted as a function of the channel length for each set of organic TFTs (**e**–**h**). In (**a**,**e**), bottom-gate, top-contact DNTT TFTs are investigated. In (**b**–**d**) and (**f**–**h**), bottom-gate, bottom-contact pentacene and C_{60} TFTs with Au contacts functionalized with 2-phenylpyrimidine-5-thiol (BP0-down), 4-(2-mercaptophenyl)pyrimidine (BP0-up), or biphenyl-4-thiol (BP0) are analyzed. For all TFTs, despite different TFT architectures, different organic semiconductors, and different contact materials, a clear channel-length dependence of the Ohmic component of the contact resistance $r_{C,0}$ is observed. This leads to a failure of the transistor model in all cases. The substantial fluctuations of $r_{C,0}$ in the short-channel-length C_{60} TFTs in (**g**,**h**) (including transistors with $r_{C,0} = 0$) reflect the fact that the uncertainty of the Ohmic contact resistances for these small channel lengths is on the order of the actual value. This large uncertainty does not obscure the clear increase of the contact resistance $r_{C,0}$ with the channel length L. The failure appears to occur for the same reason as seen for the bottom-gate, bottom-contact DNTT TFTs, because the symptom of an incorrect spacing in the saturation regime of the output characteristics is present here as well.

4. Summary and Conclusions

We have proposed a two-step fitting approach (TSFA) to check whether or not a transistor model is capable of describing the experimentally-obtained current-voltage characteristic of organic TFTs. Only a valid transistor model that correctly discriminates between contact and channel properties enables the user to reliably extract, interpret, and compare contact resistances and charge-carrier mobilities of organic TFTs. Our TSFA relies on a series of transistors with different channel lengths and consists of two steps. First, the chosen transistor model is fitted to all data points of the measured output and transfer characteristics of each TFT individually in order to extract the transistor parameters of each device. Second, it is determined whether or not the extracted parameters depend on the channel length. This consistency check is successful if (i) the measured current-voltage curves are represented well by the current-voltage curves calculated using the model and the extracted transistor parameters and if (ii) the extracted parameters are independent of the channel length. Our approach offers a clear benefit compared to conventional parameter-extraction methods in that the reliability of the underlying transistor model can be easily validated. Due to the analysis of each individual TFT as a whole, the reason for a failure of the transistor model can be identified based on the nature of the deviations between the calculated and the measured current-voltage characteristics.

We have outlined the indicators that are available to judge the consistency within the TSFA by using the transistor model underlying the conventional transmission line method (TLM) as an illustrative example. Organic TFTs with particularly small contact resistances whose operation resembles the ideal transistor behavior most closely were analyzed in particular detail. With this set of TFTs, we have demonstrated that not all types of inconsistencies can be spotted within the parameter-extraction step, but rather require a second step for validity checking. A conventional TLM analysis of this set of TFTs initially gave an apparently consistent picture comprising (i) a linear onset of the output characteristics for zero drain-source voltage and (ii) a good agreement of all calculated linear fits. However, the output characteristics calculated using the extracted parameters failed to reproduce the measured output characteristics. The subsequent validity check of the TSFA for the model underlying the TLM was not passed, because the extracted contact resistances retained a pronounced dependence on the channel length. Such inconsistencies ought to be removed or at least alleviated by improving the underlying transistor model. For example, the model underlying the TLM can be improved by accounting for a field- and charge-carrier-density-dependent mobility [7,24–26]. Even though the TSFA leads to better agreement between the calculated and the measured output characteristics, the improved model still fails the subsequent validity check of the TSFA, due to a remnant channel-length dependence of the contact resistance. The failure of the advanced transistor model featuring a field- and charge-carrier-density-dependent mobility has been demonstrated for a wide variety of organic TFTs based on different device architectures, different organic semiconductors, and source and drain contacts with poor injection behavior that caused profound non-linear contributions to the contact resistance.

To improve the currently-available transistor models, we need to face two aspects: On the one hand, the analysis of the deviations between the calculated and the measured current-voltage characteristics suggests that the charge-carrier-density dependence of the charge-carrier mobility is not properly captured. Hence, a mobility model that is particularly suitable for the predominantly two-dimensional charge transport through the charge-accumulation channel of a TFT needs to be developed. On the other hand, the gradual channel approximation should be reconsidered by accounting for the charge accumulation in the channel. Due to the lack of compensation by charges of opposite polarity within the organic semiconducting layer, this uncompensated charge accumulation affects the electric-field distribution, causing notable changes at the contacts. These changes become more important for larger channel lengths. Our TSFA can be used to check each stage of model improvement.

Supplementary Materials: The following are available online at http://www.mdpi.com/2073-4352/9/2/85/s1.

Author Contributions: Data analysis and writing, original draft preparation, M.K. and K.Z.; fabrication and measurement of DNTT TFTs, J.W.B. and H.K.; fabrication and measurement of pentacene and C_{60} TFTs, A.P., E.K.-P., G.S., and B.S.; editing J.W.B., H.K., M.K., and K.Z.

Funding: This research was funded by FWF Grant Number I 2081-N20.

Conflicts of Interest: The authors declare no conflict of interest.

References

1. Guo, X.; Xu, Y.; Ogier, S.; Ng, T.N.; Caironi, M.; Perinot, A.; Li, L.; Zhao, J.; Tang, W.; Sporea, R.A.; et al. Current Status and Opportunities of Organic Thin-Film Transistor Technologies. *IEEE Trans. Electron Devices* **2017**, *64*, 1906–1921. doi:10.1109/TED.2017.2677086. [CrossRef]
2. Paterson, A.F.; Singh, S.; Fallon, K.J.; Hodsden, T.; Han, Y.; Schroeder, B.C.; Bronstein, H.; Heeney, M.; McCulloch, I.; Anthopoulos, T.D. Recent Progress in High-Mobility Organic Transistors: A Reality Check. *Adv. Mater.* **2018**, *30*, 1801079. doi:10.1002/adma.201801079. [CrossRef] [PubMed]
3. Yamamura, A.; Watanabe, S.; Uno, M.; Mitani, M.; Mitsui, C.; Tsurumi, J.; Isahaya, N.; Kanaoka, Y.; Okamoto, T.; Takeya, J. Wafer-scale, layer-controlled organic single crystals for high-speed circuit operation. *Sci. Adv.* **2018**, *4*, eaao5758. doi:10.1126/sciadv.aao5758. [CrossRef] [PubMed]
4. Ogier, S.D.; Matsui, H.; Feng, L.; Simms, M.; Mashayekhi, M.; Carrabina, J.; Terés, L.; Tokito, S. Uniform, high performance, solution processed organic thin-film transistors integrated in 1 MHz frequency ring oscillators. *Org. Electron.* **2018**, *54*, 40–47. doi:10.1016/j.orgel.2017.12.005. [CrossRef]

5. Pasveer, W.F.; Cottaar, J.; Tanase, C.; Coehoorn, R.; Bobbert, P.A.; Blom, P.W.M.; de Leeuw, D.M.; Michels, M.A.J. Unified Description of Charge-Carrier Mobilities in Disordered Semiconducting Polymers. *Phys. Rev. Lett.* **2005**, *94*. doi:10.1103/PhysRevLett.94.206601. [CrossRef] [PubMed]
6. Li, J.; Ou-Yang, W.; Weis, M. Electric-field enhanced thermionic emission model for carrier injection mechanism of organic field-effect transistors: Understanding of contact resistance. *J. Phys. D Appl. Phys.* **2017**, *50*, 035101. doi:10.1088/1361-6463/aa4e95. [CrossRef]
7. Locci, S.; Morana, M.; Orgiu, E.; Bonfiglio, A.; Lugli, P. Modeling of Short-Channel Effects in Organic Thin-Film Transistors. *IEEE Trans. Electron Devices* **2008**, *55*, 2561–2567. doi:10.1109/TED.2008.2003022. [CrossRef]
8. Marinov, O.; Deen, M.J.; Zschieschang, U.; Klauk, H. Organic Thin-Film Transistors: Part I—Compact DC Modeling. *IEEE Trans. Electron Devices* **2009**, *56*, 2952–2961. doi:10.1109/TED.2009.2033308. [CrossRef]
9. Di Pietro, R.; Venkateshvaran, D.; Klug, A.; List-Kratochvil, E.J.W.; Facchetti, A.; Sirringhaus, H.; Neher, D. Simultaneous extraction of charge density dependent mobility and variable contact resistance from thin film transistors. *Appl. Phys. Lett.* **2014**, *104*, 193501. doi:10.1063/1.4876057. [CrossRef]
10. Natali, D.; Caironi, M. Charge injection in solution-processed organic field-effect transistors: Physics, models and characterization methods. *Adv. Mater.* **2012**, *24*, 1357–1387. doi:10.1002/adma.201104206. [CrossRef]
11. Wang, S.D.; Yan, Y.; Tsukagoshi, K. Transition-Voltage Method for Estimating Contact Resistance in Organic Thin-Film Transistors. *IEEE Electron Device Lett.* **2010**, *31*, 509–511. doi:10.1109/LED.2010.2044137. [CrossRef]
12. Takagaki, S.; Yamada, H.; Noda, K. Extraction of contact resistance and channel parameters from the electrical characteristics of a single bottom-gate/top-contact organic transistor. *Jpn. J. Appl. Phys.* **2016**, *55*, 03DC07. doi:10.7567/JJAP.55.03DC07. [CrossRef]
13. Torricelli, F.; Ghittorelli, M.; Colalongo, L.; Kovacs-Vajna, Z.M. Single-transistor method for the extraction of the contact and channel resistances in organic field-effect transistors. *Appl. Phys. Lett.* **2014**, *104*, 093303. doi:10.1063/1.4868042. [CrossRef]
14. Kanicki, J.; Libsch, F.R.; Griffith, J.; Polastre, R. Performance of thin hydrogenated amorphous silicon thin-film transistors. *J. Appl. Phys.* **1991**, *69*, 2339–2345. doi:10.1063/1.348716. [CrossRef]
15. Luan, S.; Neudeck, G.W. An experimental study of the source/drain parasitic resistance effects in amorphous silicon thin film transistors. *J. Appl. Phys.* **1992**, *72*, 766–772. doi:10.1063/1.351809. [CrossRef]
16. Natali, D.; Fumagalli, L.; Sampietro, M. Modeling of organic thin film transistors: Effect of contact resistances. *J. Appl. Phys.* **2007**, *101*, 014501. doi:10.1063/1.2402349. [CrossRef]
17. Deen, M.J.; Marinov, O.; Zschieschang, U.; Klauk, H. Organic Thin-Film Transistors: Part II—Parameter Extraction. *IEEE Trans. Electron Devices* **2009**, *56*, 2962–2968. doi:10.1109/TED.2009.2033309. [CrossRef]
18. Fischer, A.; Zündorf, H.; Kaschura, F.; Widmer, J.; Leo, K.; Kraft, U.; Klauk, H. Nonlinear Contact Effects in Staggered Thin-Film Transistors. *Phys. Rev. Appl.* **2017**, *8*, 054012. doi:10.1103/PhysRevApplied.8.054012. [CrossRef]
19. Mayer, J.; Khairy, K.; Howard, J. Drawing an elephant with four complex parameters. *Am. J. Phys.* **2010**, *78*, 648–649. doi:10.1119/1.3254017. [CrossRef]
20. Gundlach, D.J.; Royer, J.E.; Park, S.K.; Subramanian, S.; Jurchescu, O.D.; Hamadani, B.H.; Moad, A.J.; Kline, R.J.; Teague, L.C.; Kirillov, O.; et al. Contact-induced crystallinity for high-performance soluble acene-based transistors and circuits. *Nat. Mater.* **2008**, *7*, 216–221. doi:10.1038/nmat2122. [CrossRef]
21. Kraft, U.; Takimiya, K.; Kang, M.J.; Rödel, R.; Letzkus, F.; Burghartz, J.N.; Weber, E.; Klauk, H. Detailed analysis and contact properties of low-voltage organic thin-film transistors based on dinaphtho[2,3-b:2′,3′-f]thieno[3,2-b]thiophene (DNTT) and its didecyl and diphenyl derivatives. *Org. Electron.* **2016**, *35*, 33–40. doi:10.1016/j.orgel.2016.04.038. [CrossRef]
22. Petritz, A.; Krammer, M.; Sauter, E.; Gärtner, M.; Nascimbeni, G.; Schrode, B.; Fian, A.; Gold, H.; Cojocaru, A.; Karner-Petritz, E.; et al. Embedded Dipole Self-Assembled Monolayers for Contact Resistance Tuning in p-Type and n-Type Organic Thin Film Transistors and Flexible Electronic Circuits. *Adv. Funct. Mater.* **2018**, *28*, 1804462. doi:10.1002/adfm.201804462. [CrossRef]
23. Shockley, W. A Unipolar "Field-Effect" Transistor. *Proc. IRE* **1952**, *40*, 1365–1376.10.1109/JRPROC.1952.273964. [CrossRef]
24. Vissenberg, M.C.J.M.; Matters, M. Theory of the field-effect mobility in amorphous organic transistors. *Phys. Rev. B* **1998**, *57*, 12964–12967. doi:10.1103/PhysRevB.57.12964. [CrossRef]

25. Horowitz, G.; Hajlaoui, M.E.; Hajlaoui, R. Temperature and gate voltage dependence of hole mobility in polycrystalline oligothiophene thin film transistors. *J. Appl. Phys.* **2000**, *87*, 4456–4463. doi:10.1063/1.373091. [CrossRef]
26. Hall, R.B. The Poole-Frenkel effect. *Thin Solid Films* **1971**, *8*, 263–271. doi:10.1016/0040-6090(71)90018-6. [CrossRef]
27. Marquardt, D. An Algorithm for Least-Squares Estimation of Nonlinear Parameters. *J. Soc. Ind. Appl. Math.* **1963**, *11*, 431–441. doi:10.1137/0111030. [CrossRef]
28. Hong, J.P.; Park, A.Y.; Lee, S.; Kang, J.; Shin, N.; Yoon, D.Y. Tuning of Ag work functions by self-assembled monolayers of aromatic thiols for an efficient hole injection for solution processed triisopropylsilylethynyl pentacene organic thin film transistors. *Appl. Phys. Lett.* **2008**, *92*, 143311. doi:10.1063/1.2907691. [CrossRef]
29. Borchert, J.W.; Peng, B.; Letzkus, F.; Burghartz, J.N.; Chan, P.K.L.; Zojer, K.; Ludwigs, S.; Klauk, H. Small contact resistance and high-frequency operation of flexible, low-voltage, inverted coplanar organic transistors. *Nat. Commun.* **2019**, submitted.
30. Sze, S.M.; Ng, K.K. *Physics of Semiconductor Devices*; OCLC: 488586029; John Wiley & Sons: New York, NY, USA, 2007.
31. Rödel, R.; Letzkus, F.; Zaki, T.; Burghartz, J.N.; Kraft, U.; Zschieschang, U.; Kern, K.; Klauk, H. Contact properties of high-mobility, air-stable, low-voltage organic n-channel thin-film transistors based on a naphthalene tetracarboxylic diimide. *Appl. Phys. Lett.* **2013**, *102*, 233303. doi:10.1063/1.4811127. [CrossRef]
32. Bittle, E.G.; Basham, J.I.; Jackson, T.N.; Jurchescu, O.D.; Gundlach, D.J. Mobility overestimation due to gated contacts in organic field-effect transistors. *Nat. Commun.* **2016**, *7*, 10908. doi:10.1038/ncomms10908. [CrossRef] [PubMed]
33. Uemura, T.; Rolin, C.; Ke, T.H.; Fesenko, P.; Genoe, J.; Heremans, P.; Takeya, J. On the Extraction of Charge Carrier Mobility in High-Mobility Organic Transistors. *Adv. Mater.* **2016**, *28*, 151–155. doi:10.1002/adma.201503133. [CrossRef] [PubMed]
34. Fishchuk, I.I.; Arkhipov, V.I.; Kadashchuk, A.; Heremans, P.; Bässler, H. Analytic model of hopping mobility at large charge carrier concentrations in disordered organic semiconductors: Polarons versus bare charge carriers. *Phys. Rev. B* **2007**, *76*, 045210. doi:10.1103/PhysRevB.76.045210. [CrossRef]

© 2019 by the authors. Licensee MDPI, Basel, Switzerland. This article is an open access article distributed under the terms and conditions of the Creative Commons Attribution (CC BY) license (http://creativecommons.org/licenses/by/4.0/).

Article

Study on Correlation between Structural and Electronic Properties of Fluorinated Oligothiophenes Transistors by Controlling Film Thickness

Jui-Fen Chang [1,*], Hua-Shiuan Shie [1], Yaw-Wen Yang [2,3] and Chia-Hsin Wang [2]

1 Department of Optics and Photonics, National Central University, Zhongli 320, Taiwan; m222509443@yahoo.com.tw
2 National Synchrotron Radiation Research Center, Hsinchu 300, Taiwan; yang@nsrrc.org.tw (Y.-W.Y.); wang.ch@nsrrc.org.tw (C.-H.W.)
3 Department of Chemistry, National Tsing-Hua University, Hsinchu 300, Taiwan
* Correspondence: jfchang@dop.ncu.edu.tw; Tel.: +886-3-4227151 (ext. 65263)

Received: 1 February 2019; Accepted: 7 March 2019; Published: 12 March 2019

Abstract: α,ω-diperfluorohexylquaterthiophene (DFH-4T) has been an attractive n-type material employed in the development of high-mobility organic field-effect transistors. This paper presents a systematic study of the relationship between DFH-4T transistor performance and film structure properties as controlled by deposited thickness. When the DFH-4T thickness increases from 8 nm to 80 nm, the room-temperature field-effect mobility increases monotonically from 0.01 to 1 $cm^2 \cdot V^{-1} \cdot s^{-1}$, while the threshold voltage shows a different trend of first decrease then increase. The morphology of thin films revealed by atomic force microscopy shows a dramatic change from multilayered terrace to stacked rod like structures as the film thickness is increased. Yet the crystallite structure and the orientation of molecular constituent, as determined by X-ray diffraction and near-edge X-ray absorption fine structure respectively, do not differ much with respect to film thickness increase. Further analyses of low-temperature transport measurements with mobility-edge model demonstrate that the electronic states of DFH-4T transistors are mainly determined by the film continuity and crystallinity of the bottom multilayered terrace. Moreover, the capacitance-voltage measurements of DFH-4T metal-insulator-semiconductor diodes demonstrate a morphological dependence of charge injection from top contacts, which well explains the variation of threshold voltage with thickness. The overall study provides a deeper understanding of microstructural and molecular growth of DFH-4T film and clarify the structural effects on charge transport and injection for implementation of high-mobility top-contact transistors.

Keywords: organic film growth; organic transistor; charge transport and injection mechanisms

1. Introduction

In recent years, extensive studies on chemical synthesis and device engineering of n-type organic semiconductors have significantly increased the material species and electron mobilities approaching the level of p-type counterparts and have broadened the potential applications of organic field-effect transistors (OFETs) in the fields of complementary circuits and light-emitting technologies [1–3]. In addition to the derivatives synthesized based on classic fullerene [4,5], naphthalene diimide [6,7] and perylene diimide [8], many new compounds with high electron mobilities, such as tetraazapentacene-based [9] and dicyanomethylene-substituted thienoquinoidal molecules [10], have been proposed. Perfluorohexyl-substituted oligothiophene is another type of n-type molecule that has attracted much interest due to high mobility, solution processability and high thermal stability and volatility [11]. In particular, α,ω-diperfluorohexyl-quaterthiophene (DFH-4T), with the highest electron mobility among the DFH-nT (n = 2–6) series, has been intensively studied in single-component transistors [12,13] and employed as the n-channel material to realize multilayered light-emitting transistors with high ambipolar conductivity and luminescent efficiency [14,15]. Our recent study also demonstrated a high-performance ambipolar OFETs based on a combination of DFH-4T and a p-type molecule, dinaphtho[2,3-b:2′,3′-f]thieno[3,2-b]thiophene (DNTT), with matched electron and hole mobilities of ~1 cm$^2 \cdot$V$^{-1} \cdot$s^{-1} [16]. Although DFH-4T has great advantages for applications of single- and multi-component transistors, its film structure plays a critical role in determining device performance. Since DFH-4T-based transistors are generally constructed with a bottom-gate configuration, choosing appropriate substrates that support initial film growth with in-plane π–π intermolecular core interactions and low-density microstructural defects is a prerequisite for achieving high mobility. Previous work by Dholakia et al. has investigated the DFH-4T film growth on Au contact and SiO$_2$ dielectric relevant to bottom-contact transistors [17]. Interestingly, DFH-4T was found to exhibit a unique growth mode on Au surface as driven by strong chemisorption due to interfacial interactions between low-lying π* orbitals and filled metal d bands. The strong chemisorption at DFH-4T/Au surface and the DFH-4T intermolecular interactions in the bulk film derive a significant transition in molecular orientation and morphology during growth, unlike the fixed molecular orientation as film grows on SiO$_2$. For bottom-contact DFH-4T transistors, the different growth modes on Au contacts and SiO$_2$ dielectric can result in the formation of inhomogeneous microstructures unfavorable for electron injection and therefore mobilities are typically very low (<10^{-4} cm$^2 \cdot$V$^{-1} \cdot$s^{-1}). By contrast, top-contact transistors can yield much higher mobilities and has been commonly adopted in multi-component device applications. When it is necessary to deposit metal contact or even organic semiconductor on top of DFH-4T film, surface morphology is another important feature to be concerned, since it directly affects the quality of heterointerface that determines charge injection from top contacts or transport property of organic semiconductor deposited thereon and can lead to very different OFET function [16]. DFH-4T film morphology depends on manifold factors such as the underlying substrate, deposition conditions and thickness and usually exhibits complex microstructures. In order to optimize the top-contact DFH-4T transistors and broaden their related applications, more detailed studies are needed to understand how the DFH-4T film grows from interface to bulk on a given substrate and which structural factors can essentially influence the transport and injection properties.

Previously, we reported the highest mobility of 2.1 cm$^2 \cdot$V$^{-1} \cdot$s^{-1} for a single-component DFH-4T transistor with the top-contact geometry and suggested that the morphological feature with randomly stacked large-scale rod like crystals is a key attribute for effective charge injection and superior device performance [16]. Behind this complex film morphology, however, how the structure is formed and its effect on the charge transport and injection mechanisms remain unclear. Based on the same processing conditions, here we take a close examination of the structural properties of DFH-4T film with thickness ranging from monolayer to several tens of nanometers and correlate the structure property change with the charge transport and injection characteristics of the top-contact OFET. The paper is organized as follows. The transistor characterizations as a function of DFH-4T thickness is reported in Section 2. In Section 3 we report on the characterization of the thickness-dependent thin film properties such as morphology, microcrystalline state and orientation of molecular constituents by mean of atomic force microscopy (AFM), X-ray diffraction (XRD) and near-edge X-ray absorption fine structure (NEXAFS) measurements, respectively. In Section 4 we perform the low temperature transport measurement and analysis with mobility-edge (ME) model to study the electronic states of DFH-4T OFETs with different thicknesses. This allows us to clarify the microstructural factors that determine the transport property of DFH-4T films. In Section 5 we discuss the morphological effects on the charge injection of DFH-4T OFETs on the basis of capacitance-voltage (C–V) response measurement. A conclusion is given in Section 6.

2. Transistor Characterization

The DFH-4T transistors for this study were fabricated in a bottom-gate, top-contact configuration (Figure 1a). We used a heavily n-doped Si wafer with 300 nm thermally grown SiO$_2$ as the substrate. After cleaning the substrate, 200 nm thick poly (methyl methacrylate) (PMMA) was spun on the Si/SiO$_2$ substrate to facilitate the microstructural growth of DFH-4T and provide a good dielectric with few –OH groups that could trap electrons. This yields a total insulator capacitance $C_i = 6.4$ nF·cm^{-2} for the transistors. The DFH-4T (Lumtec Co) was then deposited onto the substrate at 323 K in high vacuum (<10^{-6} mbar) at a rate of 0.1 Å/s, with the thickness varied from 3 nm to 80 nm. Final thicknesses were monitored by a quartz crystal microbalance during deposition process and calibrated by using AFM after deposition. Hereafter, the film thicknesses are described as the mean thickness value. Finally, the top source/drain electrodes (80 nm Ag) was evaporated on the DFH-4T layer to complete the transistor. The channel length (L) and width (W) were 100 µm and 1500 µm, respectively. All the measurements of I–V characteristics of transistors were carried out in a Janis ST300 cryostat (Janis Research Company, Woburn, MA, USA) at a controlled temperature from 300 K to 80 K with Keysight B1500A semiconductor parameter analyzer (Keysight Technologies, Santa Rosa, CA, USA).

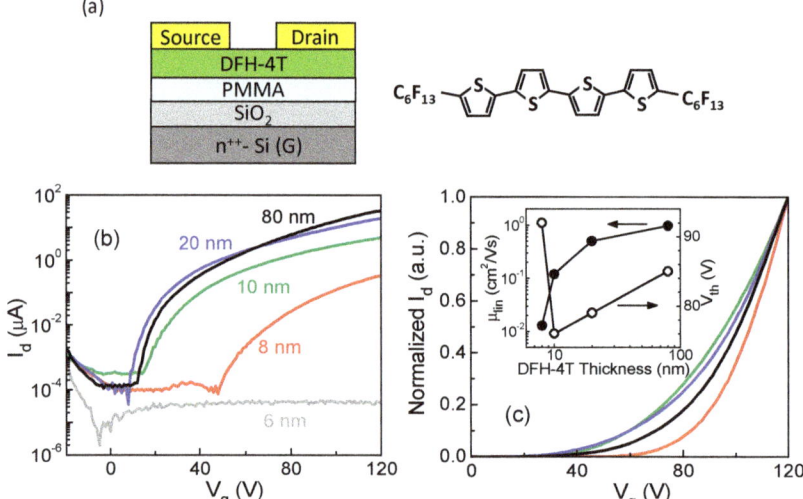

Figure 1. (**a**) Chemical structure of DFH-4T and schematic illustration of the bottom-gate, top-contact DFH-4T organic field effect transistor (OFET). (**b**) Linear transfer characteristics ($V_d = 10$ V) of DFH-4T transistors with various DFH-4T thicknesses at room temperature (300 K). (**c**) The same transfer characteristics as shown in (**b**) plotted on a linear scale and normalized respectively with the maximum current at $V_g = 120$ V, revealing different turn-on thresholds of various devices. The inset shows the extracted linear mobility and threshold voltage as a function of DFH-4T thickness.

Figure 1b shows the linear transfer characteristics ($V_d = 10$ V) of DFH-4T transistors with different thicknesses at room temperature (300 K). No conductive current was detected in the device with 6 nm (and thinner) DFH-4T. The transistor behavior appears when the DFH-4T thickness increases to 8 nm but a high onset voltage V_{on} (~50 V) and low on-current imply the presence of a large number of electronic defect states. For the 10 nm device, V_{on} is significantly reduced to less than 20 V and the on-current is increased by more than one order of magnitude as compared to the 8 nm device. When the thickness exceeds 10 nm, the device shows no apparent shift of V_{on}. The on-current is increased by few folds between 10 nm and 20 nm but only increased a little between 20 nm and 80 nm. Figure 1c shows the same transfer curves normalized respectively with their maximum current at $V_g = 120$ V, revealing more clearly the turn-on behavior of different devices. We calculate the linear field-effect mobilities (μ_{lin}) by using a conventional equation

$$\mu_{lin} = \frac{L}{WC_iV_d} \frac{I_d}{(V_g - V_{th})} \quad (1)$$

where V_{th} is the threshold voltage extracted by extrapolating the linear portion of the I_d versus V_g curve to $I_d = 0$. The inset of Figure 1c summarizes how μ_{lin} and V_{th} vary with the DFH-4T thickness. Overall, μ_{lin} increases sharply from 0.01 $cm^2 \cdot V^{-1} \cdot s^{-1}$ to 0.5 $cm^2 \cdot V^{-1} \cdot s^{-1}$ in the range of 8–20 nm and tends to level off above 20 nm, reaching about 1 $cm^2 \cdot V^{-1} \cdot s^{-1}$ at the 80 nm. In contrast, it is interesting to note that V_{th} is highest in the 8 nm device, then decreases abruptly in the 10 nm device and increases again above 10 nm. For OFETs, it is generally expected that V_{th} decreases with the increasing thickness for the first few monolayers due to a reduction of interface trap states and becomes roughly fixed once the film is sufficiently thick due to a stabilization of interface trap density. The nonmonotonic variation of V_{th} in DFH-4T transistors implies some underlying mechanisms that might involve electron injection and accumulation above 10 nm. As will be discussed below, the thickness dependence of DFH-4T transistor characteristics can be well understood from the structural evolution in the film growth.

3. Structural Property Characterization

AFM measurements (Veeco/DI NanoMan D3100CL, EnviroScope AFM in tapping mode) were first performed to shed light on the possible correlation between the transistor performance and the DFH-4T film morphology (Figure 2). In the 3 nm film, it can be seen that the initial growth of the first layer on PMMA exhibits isolated dendrimer microstructures. In the 6 nm film, the first layered dendrimers grow larger but remain isolated, while the second layered terrace structures begin to form. The multilayered terrace structures in the 8 nm film reach a continuum percolation threshold, so that the transistor becomes conductive. However, a prevalent presence of voids and structural defects could be the main cause for a high turn-on threshold and low mobility. It is noted that the measured step heights in the 3–8 nm films are approximately multiples of 3 nm. This corresponds to one-half of the DFH-4T *a*-axis unit cell dimension [17], implying that the long molecular axis is aligned along the surface normal. In the 10 nm film, the multilayered terrace structures cover the surface to a greater extent, which may account for the sharply increased mobility. Interestingly, a low density of large-scale rod like crystals, with a length of 100–1000 nm and a height of ~100 nm, form on top of terrace structures. For even thicker films, the density of rod like crystals is found to increase with thickness and eventually forms a densely packed crystal network in the 80 nm film. The height and diameter of the crystals also increase with thickness (~200 nm in the 80 nm film) but the lengths of the crystals do not increase much. However, the increased density and size of rod like crystals do not well correspond to the mobility variation for the thickness from 10 nm to 80 nm. As seen in the 20 nm film, the crystals are not yet formed into a network and unlikely contribute to the effective current pathways that result in a five-fold increase of mobility for the thickness from 10 nm to 20 nm. And even an interconnected crystal network is formed in the 80 nm film, the mobility shows little enhancement for the thickness between 20 nm and 80 nm. We therefore conjecture that the improved film quality of the bottom multilayered terrace in the 10–80 nm thick films may reflect more on the mobility variation. The surface coverage of the terrace structures could be almost complete in the 20 nm film but not much further improved in the 80 nm film, so that the mobility still increases for the thickness from 10 nm to 20 nm and becomes saturated for the thickness of more than 20 nm.

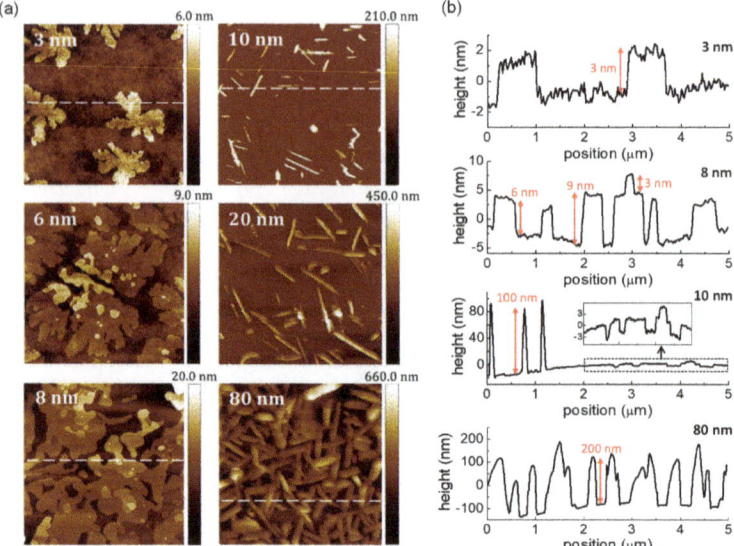

Figure 2. (a) Atomic force microscopy (AFM) topographical images (5 μm × 5 μm) of DFH-4T films with different thicknesses deposited on PMMA dielectric and (b) height profile cross section for a few selected thicknesses, indicating the formation of multilayered terrace structure below 10 nm and large rod like crystals above 10 nm. The height of each layer in the terrace structure is measured to be ~3 nm. On the other hand, the height and diameter of rod like crystals tend to increase with thickness, from ~100 nm in the 10 nm film to ~200 nm in the 80 nm film.

To assess the crystallinity of the DFH-4T films, synchrotron based XRD diffraction patterns were obtained and presented in Figure 3. DFH-4T in an all-trans configuration has a long molecular dimension of about 30 Å. For the 8 nm thick film, a set of (h00) diffraction peaks already emerge but weaker in intensity due to the presence of only a few repeated molecular units along surface normal direction. As the thickness of the film is increased, the diffraction peaks gain in intensity, as expected. For the thickest film of 80 nm, the most intense (400) peak appears at 5.80° and is accompanied by (600) peak at 8.78° as well as some even weaker but discernible high-order diffraction peaks. Diffraction order was assigned by consulting with earlier published paper [18], in which the large unit cell (a = 61.095 Å, b = 5.750 Å, c = 8.974 Å; $\alpha = \gamma = 90°$, $\beta = 94.5°$) contains four DFH-4T molecules with two DFH-4T arranged along a-axis direction. The fact that the observed diffraction peaks are mostly of (h00) type indicates that the DFH-4T crystallites are formed with their b,c axes of the unit cell in parallel with the substrate surface, resulting in a large d_{200} spacing calculated to be 60.91 Å. Thus, the DFH-4T molecules for the as-arranged crystallites tend to have their long molecular axes aligned perpendicular to the substrate, with in-plane π-π stacking favorable for charge transport. The slight discrepancy in diffraction peak position found between the thin and thick films suggests the presence of structural strain in the thin films. Overall, the thicker films have a higher degree of microstructural order and a smaller interlayer spacing. It is also noted that as the films become thicker with the formation of rod like crystals (>10 nm), new diffraction peaks emerge at higher angles, (11$\bar{1}$) at 18.44° and a smaller one (311) at 19.03°, indicating the presence of some other differently oriented crystallites in the rod like structure. Unlike (h00) crystallites, (311) and (11$\bar{1}$) crystallites tend to have their long molecular axes aligned parallel to the substrate, which may hinder the charge transport to some extent. Probably, the surface roughness of rod-like crystals and their staking geometry may not be able to support a uniform growth of DFH-4T crystallites.

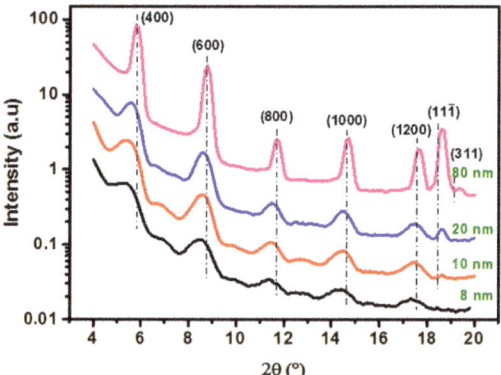

Figure 3. The θ–2θ diffraction patterns obtained with X-ray with an energy of 8.0 keV (λ = 1.5498 Å) for the DFH-4T films of different thickness deposited on PMMA dielectric. The intensity scale is given in a logarithmic scale to highlight the pattern development with thickness. The assignment of diffraction order is consistent with the structure data reported in Ref. [18]. The expected diffraction angles were marked with dash-dot lines.

As XRD is sensitive to the long-range structural ordering of materials but provides very little information about non-crystalline or amorphous part of materials, we thus use a somewhat complementary technique, namely NEXAFS, to reveal the orientation of the molecular constituents of organic materials, irrespective of whether a long-range structural ordering exists. Specifically, by exploring the polarization dependence of NEXAFS π-resonance signal, one can learn how the quarterthiophene units of DFH-4T molecules orient themselves on the surface. This piece of information is important because the degree of resultant π–π overlap greatly influences the charge transport in OFET. The measurements were carried out in a UHV surface science endstation located at the beamline BL24A of Taiwan Light Source (TLS) of NSRRC. NEXAFS data were acquired by means of a homemade partial electron yield (PEY) detector comprising a set of retarding meshes in conjunction with an electron amplification device of microchannel plate. For carbon K-edge spectral acquisition, the retarding voltage was set at −150 V to retard all but the signals derived from carbon Auger decay process for an enhanced signal to background ratio. Figure 4a shows the relevant angular coordinates in measurements, where the soft X-ray with polarization factor (P) determined to be 0.85 for ~300 eV photons is incident at an angle of θ from the sample surface plane. The NEXAFS spectra were then obtained at various values of θ to obtain the angular dependence of the X-ray absorbance. Figure 4b,c show a set of edge-jump normalized carbon K-edge NEXAFS spectra taken at different X-ray incident angles for the 8 nm and 80 nm DFH-4T films grown on PMMA/SiO$_2$. The first major X-ray absorption feature before 286.3 eV is attributed to the 1s-π* resonance and the second absorption feature (286.3 eV to 288.0 eV) is derived from a mixed σ* resonances of C–H and C–S types [19,20]. The third absorption feature at 288.5 eV is due to the π* (C=O) resonance of PMMA [21] and becomes much reduced in intensity for 80 nm thick DFH-4T film. The observation of C=O resonance indicates an incomplete covering of PMMA by DFH-4T because otherwise no PMMA signal is expected given a NEXAFS probing depth of 6 nm for the present PEY detection with 150 V retardation voltage [22]. The introduction of two perfluorinated side chains in DFH-4T changes the usual featureless σ* resonance starting from ~290 eV into pronounced peaks. Previous NEXAFS study of oriented perfluoroeicosane (CF$_3$(CF$_2$)$_{18}$CF$_3$) film showed that the absorption onset was shifted from ~284 eV to 290 eV due to the strong chemical shift induced by C–F bond and three prominent σ* resonances of C–C and C–F types appear between 290 eV and 305 eV [23]. Based on this work, the three major features in our spectra are assigned as follows: σ* (C–F) at 292.6 eV, σ* (C–C) at 295.3 eV, and σ* (C–F) at 298.7 eV. We believe that the most reliable way of determining the tilt angle of DFH-4T is to evaluate the X-ray incidence angle dependence of the fitted intensity of low-energy 1s-π* resonances located between 282.5 eV and 286.25 eV (inset of Figure 4b,c). In this energy window, neither PMMA underlayer nor the perfluorinated side chains contributes any absorption signal and the signal itself is derived from the coplanar quarterthiophene exclusively, which greatly simplifies the data interpretation. The next high-lying absorption feature originated from C–H and C–S resonances has a less clear polarization dependence because of mutually orthogonal σ- and π-polarization dependence from the carbons of thiophenes [20,24]. The analysis based on C–F and C–C σ* resonances is seemingly appealing; however, the presence of step edge and the angle-independent background can make a reliable analysis difficult unless a more complete set of data is available [25].

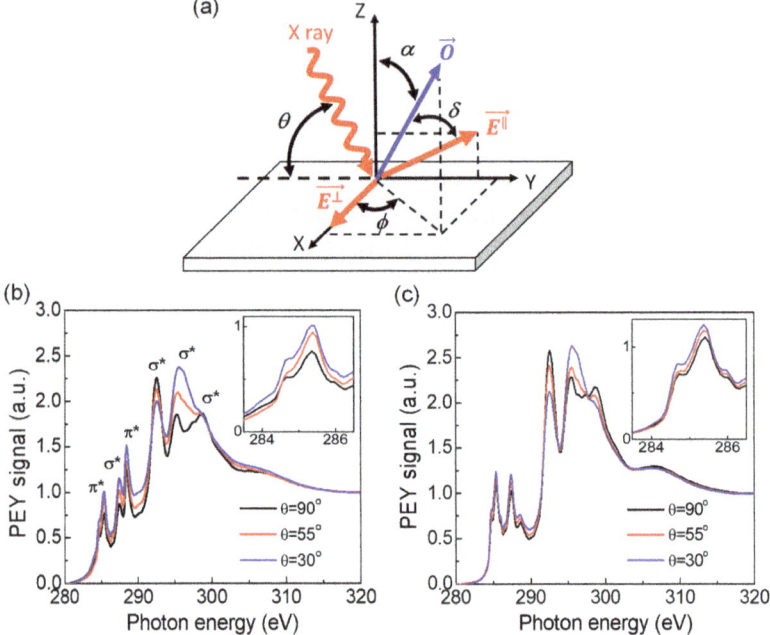

Figure 4. (a) Configuration of NEXAFS measurements. X-ray is incident in the *yz* plane and at an angle θ from the surface. The major electric field component $E^{\|}$ lies in the storage ring plane *yz* and the minor electric field component $E^{\perp}$ is in *xz* plane. Transitional dipole moment vector $\vec{O}$ is inclined at an angle α from the surface normal direction (*z*) and this inclination angle is determined with NEXAFS. In comparison, the azimuthal angle ϕ cannot be determined in most cases due to the coexistence of multiple surface domains that results in azimuthal averaging. (b) and (c) show the N K-edge partial electron yield (PEY) data for the 8 nm and 80 nm films deposited on PMMA dielectric, respectively. Scans were taken with the X-ray incident angles at θ = 30°, 55°, 90° measured from surface plane. The inset shows the enlarged plot of the first absorption feature before 286.3 eV, which is attributed to the 1s-π* resonance.

With this in mind, we carried out a complete fitting at the angular dependence of the 1s-π* resonance intensity, as shown in Supplementary Materials Figure S1. The method of estimating the tilt of aromatic ring plane based on crystallographic data was presented in the previous NEXAFS investigation of organic semiconducting film of anthradithiophene derivative [26]. For the 8 nm film, the normalized intensity values are 0.915/1.00/1.18 for 90°/55°/30° incidence angle, which gives a transition dipole momentum vector (perpendicular to thiophene ring plane) at ~50° from the surface normal. That is, the thiophene ring plane is at ~40° from the surface normal. For the 80 nm film, the normalized intensity values are 0.964/1.00/1.07 for 90°/55°/30° incident angle, indicating that the thiophene ring plane is tilted toward the surface normal by another ~2°. Interestingly, probing of the 8 nm and 80 nm films with drastically different surface morphologies (terrace vs rod like crystals) shows very similar molecular inclinations. It is noted that an averaged ~40° tilt is larger than the molecular inclination deduced from the dominant (h00) diffraction in the X-ray data, where the aromatic ring plane is estimated to be inclined from the surface normal by 30° (Figure S2). This discrepancy may arise from a fact that the XRD data are derived solely from the crystalline part of the materials while the NEXAFS data are contributed from *all* the molecules, setting a basis for the data difference. Nonetheless, the NEXAFS data are consistent with the X-ray result that DFH-4T mainly adopts an edge-on molecular orientation. From these two techniques, it can be claimed that the DFH-4T films are constituted mainly with (h00) crystallites but also with some differently oriented crystallites and amorphous components, however, the overall crystallite structure and molecular orientation do not show much difference with respect to film thickness increase.

4. Temperature Dependent Transport Measurement and Analysis

Next, we conduct the low temperature measurement to investigate the electronic states and transport properties of DFH-4T transistors with various thicknesses. Figure 5a–d show the linear transfer characteristics (V_d = 10 V) of DFH-4T transistors over a temperature range from 300 K to 100 K for different thicknesses from 8 nm to 80 nm, plotted on both a logarithmic and a linear scale. As the temperature decreases, all the devices exhibit a decrease of on-current and a positive shift of V_{on}, indicating an increased trapping of immobile charges in the deep localized states. V_{th} is shifted to more positive values at lower temperature for the thicknesses of 8–10 nm and tends to be independent of temperature for the thicknesses of >10 nm. To compare these devices with different temperature dependences of V_{th}, we include the threshold voltage effect into calculation of the gate dependent mobility by taking transconductance ($\partial I_d / \partial V_g$) of the transfer curves in the linear regime [27]

$$\mu_{lin}(V_g, T) = \frac{L}{W C_i V_d} \left(\frac{\partial I_d}{\partial V_g} \right)_{V_g - V_{th} \gg V_d} \tag{2}$$

with the fixed $V_g - V_{th}$ for all the temperatures. In this way the amount of accumulated charges in the channel is determined only by the values of $V_g - V_{th}$ and is independent of temperature. Figure 5e–h show the temperature dependence of linear mobilities extracted for different transistors. In the linear regime, the choice of $V_g - V_{th}$ values needs to be sufficiently large as compared to V_d. Note that this condition is not well satisfied for the 8 nm device, since its relatively high V_{th} at low temperature limits the maximum value of $V_g - V_{th}$ in the gate sweep range to be only slightly larger than V_d. Nevertheless, because the transconductance of the 8 nm device is weakly dependent on V_g above V_{th}, the extracted mobility could be similar to but only slightly lower than that if extracted with larger $V_g - V_{th}$ and is still valid for further analysis. For all the devices, it can be seen that the temperature dependent mobility is nonmonotonic. Below 240 K the mobilities approximately follow the Arrhenius rule. The activation energy extracted from 80–240 K is 23.4 meV for 8 nm device and reduces to 8.3 meV for 80 nm device. Above 240 K the mobilities increase with temperature in a steeper slope compared to the lower temperature regime. Such a change of slope at elevated temperature has also been observed in some microcrystalline organic semiconductors and been interpreted as due to a nonzero band-tail mobility in the mobility-edge (ME) model [28]. From the low activation energy extracted below 240 K, we speculate that the DFH-4T films have a relatively narrow distribution of trap states. Therefore, in addition to being thermally activated, small amounts of charge could tunnel through the trap states and lead to the increased slope. For simplicity, we mainly analyze the low temperature regime (80–240 K) with the ME model, where the mobility in the band-tail states can be reasonably assumed to be zero.

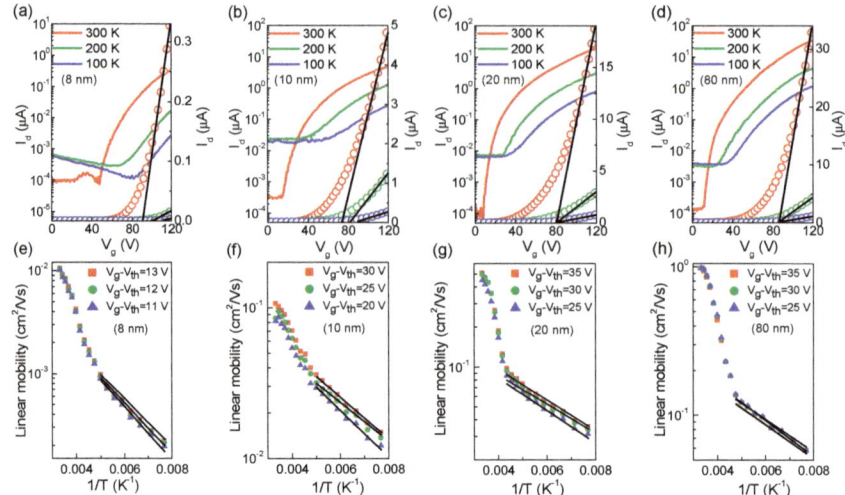

Figure 5. (**a**–**d**) Temperature dependence of the linear transfer characteristics (V_d = 10 V) of the DFH-4T transistors with various thicknesses, plotted on a logarithmic scale and a linear scale (open circles). The black lines provide a visual guide to the values of V_{th} for the different temperatures. For the transistors with thicknesses of 8–10 nm, V_{th} shifts to larger value as temperature decreases but remains the same for the greater thickness of 20–80 nm. (**e**–**h**) Temperature dependence of the linear mobilities of the DFH-4T transistors extracted from the transfer characteristics as shown in (**a**–**d**) for various thicknesses (solid dots) and the effective mobilities calculated by using the ME model (lines) are shown for comparison.

The ME model assumes that the mobile and localized states of a semiconductor are separated by a defined energy, that is, the mobility edge and carriers are transported in the mobile states with a constant mobility (μ_i) and trapped in the localized states with zero mobility. According to Ref. [29], we define the mobility edge as E = 0 eV. The localized states (E > 0) are described with an exponential density of states (DOS) that decay in the band gap,

$$D_{tail}(E) = \frac{N_{tail}}{w_{tail}} \exp\left(-\frac{E}{w_{tail}}\right), \qquad (3)$$

where N_{tail} represents the concentration of band-tail states and w_{tail} is the width of exponential trap distribution. On the other hand, the mobile states (E < 0) are described based on a simple form for 3-D free electron gas ($D(E) \sim E^{1/2}$ [30]):

$$D_{mob}(E) = \frac{N_{tail}}{w_{tail}} \sqrt{\frac{E_c - E}{E_c}}, \qquad (4)$$

where E_c is a free parameter used to tailor the shape of DOS in the mobile states. Since the variation of E_c is insensitive to the fitting result of mobilities [31], we fix E_c = 30 meV in this work. When the gate bias is applied, the total accumulated charge concentration (N_{tot}) would equal to the summation of the charge concentrations in the mobile states and in the localized states:

$$N_{tot} = \frac{C_i|V_g - V_{th}|}{h} = \int_{-\infty}^{0} D_{mob}(E)f(E_F, E)dE + \int_{0}^{\infty} D_{tail}(E)f(E_F, E)dE, \qquad (5)$$

where h = 1 × 10^{-7} cm is the channel dimension normal to the dielectric surface and f(E$_F$, E) is the Fermi-Dirac distribution of electrons. After solving the Fermi energy E$_F$(V$_g$, T) in Equation (5), the charge concentrations in the mobile states

$$N_{mob} = \int_{-\infty}^{0} D_{mob}(E) f(E_F, E) dE \qquad (6)$$

can be calculated and then the effective mobility can be determined by the fraction of mobile charges,

$$\mu(V_g, T) = \mu_i \frac{N_{mob}(V_g, T)}{N_{tot}}. \qquad (7)$$

In Figure 5e–h we show the best fit to the temperature dependent mobility at different V$_g$–V$_{th}$ values for all the devices by using Equation (7) with μ_i, N$_{tail}$ and w$_{tail}$ as free parameters. The values of the fitting parameters are given in Table 1. Overall, the transistors with thicker DFH-4T have higher intrinsic mobilities μ_i and narrower bandwidths w$_{tail}$ and lower concentrations N$_{tail}$ of localized states. From the above structural characterizations, it is not surprising that a high population of film voids, discontinuities and a high degree of microstructural disorder in the 8 nm device result in a high energetic disorder and hence the largest N$_{tail}$ and w$_{tail}$. On the other hand, the intrinsic mobility reflects the degree of polaron delocalization in the mobile states [28]. The poorer crystallinity of the 8 nm device may lead to an increased localization of the polaron wave function in the crystalline domains, which accounts for a low band mobility. As the film grows thicker, the film continuity and crystallinity are improved, so that the energetic disorder decreases, and intrinsic mobility increases. Importantly, the ME model predicts that the increase in μ_i and decrease in w$_{tail}$ and N$_{tail}$ are most pronounced for the thickness of 8-10 nm but not obvious for the thickness exceeding 20 nm. This result provides unambiguous evidence that the electronic state of DFH-4T transistors is mainly related to the structural properties of bottom terrace but is weakly related to the conformation of top rod like crystals. Therefore, we can confirm that the conductive channel is located at the bottom terrace, consistent with the general expectation that gate-induced charge accumulation occurs within a few nanometers from the dielectric/semiconductor interface. The high mobility thus relies primarily on reduction of trap states by improving film continuity and crystalline quality of the terrace structure.

Table 1. Parameters extracted from fitting the experimental data to the ME model.

Thickness (nm)	μ_i (cm$^2 \cdot$V$^{-1} \cdot$s^{-1})	w$_{tail}$ (meV)	N$_{tail}$ (cm^{-3})
8	0.0063	20	5.3 × 10^{19}
10	0.11	16	4.8 × 10^{19}
20	0.207	14.8	4.7 × 10^{19}
80	0.34	14.3	4.3 × 10^{19}

5. Morphological Effect on Charge Injection

Although the conformation of top rod like crystals is shown to be little correlated with the charge transport, how these crystals influence the charge injection from the top electrodes remains a question. To evaluate this effect on the transistor performance, we measure the low-frequency C–V response of the metal-insulator-semiconductor (MIS) diodes (Si/SiO$_2$/PMMA/DFH-4T/Ag) of varied DFH-4T thickness. The metal-insulator-metal (MIM) diodes (Si/SiO$_2$/PMMA/Ag) were also measured as a control. In the MIS diodes the DFH-4T and top Ag electrode were sequentially evaporated through a fixed shadow mask to eliminate the parasitic capacitance. The same shadow mask was also used to define the top contact of the MIM diode. The C–V measurements were performed using a 4192 A Hewlett Packard Impedance analyzer in a nitrogen glove box. Since the maximum output voltage of impedance analyzer is limited at 35 V, we fabricated the MIS and MIM diodes with the thinner dielectrics (70 nm SiO$_2$ and 90 nm PMMA) to yield the dielectric capacitance (19.4 nF·cm^{-2}) about three times higher than that of transistors (6.4 nF·cm^{-2}). In this way, a complete switch-on behavior of the MIS diodes can be obtained with an applied voltage of 35 V, which corresponds to that of transistors as driven with $V_g \sim 100$ V.

Typically, for a planar MIS diode, the capacitance has the lowest value in depletion and starts to increase with V_g above the threshold as the accumulated charges move closer to the semiconductor/insulator interface and saturates at the value approaching the insulator capacitance in the high V_g regime. That is, the C–V response also indicates how the resistance (including contact and semiconductor bulk) varies with V_g. As the result shown in Figure 6a, all the DFH-4T MIS diodes exhibit the C–V response in a similar manner as expected for typical planar diodes but the values of capacitance in depletion and accumulation regimes strongly depend on the thickness. In depletion regime the capacitance decreases as the thickness increases, which is as expected because the thicker DFH-4T film has a lower capacitance and hence a lower serial capacitance of the insulator and semiconductor capacitors. The charge injection in different MIS diodes manifests itself more clearly in accumulation regime. Opposite to the trend in depletion regime, the capacitance is found to increase with the thickness in accumulation regime. In the 8 nm diode, the capacitance saturates at a relatively low value compared to the insulator capacitance measured from the MIM diode (C_{in}), indicating the presence of a portion of nonconductive region due to a high population of structural defects in the multilayered terrace. The capacitance of the 10 nm diode saturates at a higher value and approaches that of MIM diode, suggesting an improved structural quality. Interestingly, the capacitance of the 20 nm diode exceeds that of MIM diode in the high V_g regime. This may be resulted from two phenomena: First, the nonconductive defects almost diminish in the multilayered terrace at the thickness between 10 nm and 20 nm, as evidenced by the AFM measurement (see Figure 2). Second, the topography of top rod like crystals in the 20 nm DFH-4T could be sufficiently rough, so that the effective contact area (A_{eff}) of DFH-4T/metal interface is apparently larger than the contact area (A_{in}) of MIM diode, resulting in a higher effective capacitance $C_{eff} = C_{in}A_{eff}/A_{in}$ [16]. In the 80 nm diode, the capacitance undergoes an obvious "two-stage" increase from the lowest value in depletion regime to the highest value in accumulation regime, suggesting a nonmonotonic charge accumulation process. As illustrated in Figure 6b, in the first stage (low positive V) the injected charges fill the crystal boundary defect states and disorder sites within the crystals, forming conducting pathways between the top contacts and the bottom terrace. Increase in V_g allows injection of more charges and leads to the formation of more conducting pathways and enhances conductivity of crystal network, so that the capacitance increases accordingly. In this stage, only the top crystal network becomes conducting whereas the bottom terrace remains insulating. The inflection point between the first and second stages (capacitance approaching that of thinner MIS diodes in depletion regime, $\sim$18.8 nF·cm^{-2}) occurs when the charges start to be accumulated in the bottom terrace, which corresponds to the effective turn-on threshold of transistor characteristic. In the second stage (high positive V), the capacitance continuously increases with V_g and reaches an even higher value than that of the 20 nm diode without apparent saturation. Again, this can be interpreted with the topography of top rod like crystals in the 80 nm DFH-4T. Due to a very large A_{eff} of DFH-4T/metal interface in the crystal network, injection of more charges with higher V_g produces even more available conducting pathways. Therefore, not only the conductivity of the crystal network continuously increases but more charges can be accumulated in the bottom terrace, which together leads to an unsaturated and a very high C_{eff} in the high V_g regime. Overall, the C–V measurement provides a clear insight of how the conformation of top crystals affects charge injection and accumulation in the DFH-4T transistors. The thicker films with more densely packed crystals on one hand have more defect states that need to be filled before effective accumulation takes place in the bottom terrace, while on the other hand can produce a larger A_{eff} of DFH-4T/metal interface that allows injection of more charges to contribute a higher current. This well explains why the transistors with >10 nm DFH-4T exhibit an unexpectedly increased V_{th} (Figure 1c) but also a sharply increased on-current, which is particularly evident in the 80 nm device. Such a morphological effect on charge injection/accumulation processes should also be taken into account in some small molecule devices with similar large crystal growth at sufficiently large thicknesses [32]. We speculate that the formation of rod like crystals in the DFH-4T films of >10 nm might be resulted from several mechanisms. First, a small amount of aggregates formed in the first few layers of terrace structure serve as the seeds to initiate the crystal growth. Second, due to the high electronegativity of fluorinated side chains, the dipole-dipole interaction between molecules could be so strong to hold molecules together, resulting in a rapid growth of large crystals into three dimensions. Third, the π-stacking interaction between thiophene rings may favor elongation of crystals in in-plane direction. By developing novel deposition approaches capable of controlling various factors on the crystal growth, it is possible to minimize the crystal defects and maximize the contact area with the top electrodes, so that V_{th} can be further reduced while achieving a higher on-current.

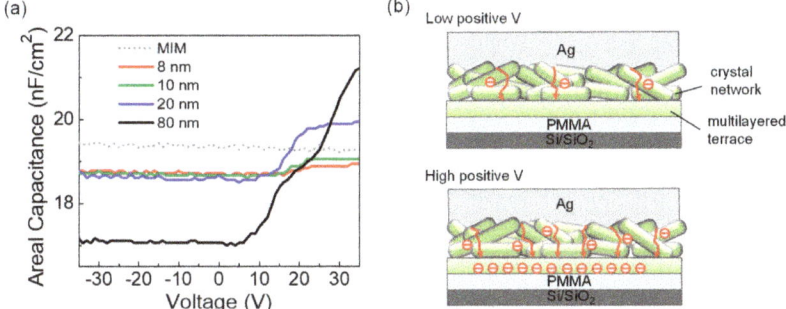

Figure 6. (a) Capacitance-voltage curves of the DFH-4T MIS diodes with various DFH-4T thicknesses and the metal-insulator-metal (MIM) diode. (b) Illustration showing the electron injection and accumulation at low and high positive applied voltage in the 80 nm DFH-4T MIS diode with the film thickness not shown to scale.

6. Conclusions

We present a detailed study of the correlation between the electronic and structural properties of the DFH-4T transistors as controlled by film thickness. The transistor characterizations show that the field-effect mobility becomes observable in the film of 8 nm, then increases sharply from 8 nm to 10–20 nm and levels off above 20 nm. The threshold voltage first decreases from 8 nm to 10 nm, then increases with thickness above 10 nm. Such variations in the mobility and threshold voltage are closely correlated with the structural evolution in the film growth. The morphology characterizations show that DFH-4T films evolve form multilayered terrace structure for the first 10–20 nm to a structure dominated with randomly distributed rod like crystals at larger film thickness. Despite this drastic change in the morphology, the molecules of the films exhibit similar orientation but are packed into the crystallites of improved ordering for the thicker films. Low temperature transport analyses further demonstrate that the electronic state and transport property of DFH-4T transistors are determined by the bottom terrace structure. The significantly improved film continuity and crystalline quality of the bottom terrace as the thickness increases from 8 nm to 20 nm result in an apparently reduced bandwidth of shallow trap states and enhanced intrinsic mobilities, which accounts for the sharply increased mobility. On the other hand, the rod like crystals appearing above 10 nm are shown to weakly affect the transport property in the channel region but have a direct impact on the charge injection and accumulation process. The C–V measurements of the DFH-4T MIS diodes clearly suggest that a higher density of rod like crystals in a thicker film has more defects in the crystal regions but also produces a larger effective DFH-4T/metal contact area for charge injection. This explains an increased threshold voltage, but higher on-current levels achieved in the transistors with thickness exceeding 10 nm. Put it all together, our study affords a better understanding of the structural factors that govern charge transport and injection in DFH-4T transistors with complex film morphology.

When processing DFH-4T films for various device applications, it is not only important to optimize the structural order of bottom terrace to achieve high electron mobilities but also necessary to control the growth of top crystals that may impose strong morphological effects on subsequently deposited metals or semiconductors and affect the overall charge injection/accumulation properties in the devices.

Supplementary Materials: The following are available online at http://www.mdpi.com/2073-4352/9/3/144/s1, Figure S1: The fitting of NEXAFS spectrum, Figure S2: Molecular constituents of (h00)-aligned DFH-4T crystallites.

Author Contributions: J.-F.C. designed the experiments and wrote the main text of the manuscript. H.-S.S. fabricated the devices and conducted AFM, temperature dependent transport and C–V measurements. Y.-W.Y. and C.-H.W. contributed to the X-ray diffraction and NEXAFS measurements and data interpretation.

Funding: This research was funded by Ministry of Science and Technology of Taiwan under Contract number MOST 105-2112-M-008-008-MY3.

Conflicts of Interest: The authors declare no conflict of interest.

References

1. Usta, H.; Facchetti, A.; Marks, T.J. n-Channel semiconductor materials design for organic complementary circuits. *Acc. Chem. Res.* **2011**, *44*, 501–510. [CrossRef]
2. Muccini, M. A bright future for organic field-effect transistors. *Nat. Mater.* **2006**, *5*, 605–613. [CrossRef]
3. Usta, H.; Sheets, W.C.; Denti, M.; Generali, G.; Capelli, R.; Lu, S.; Yu, X.; Muccini, M.; Facchetti, A. Perfluoroalkyl-functionalized thiazole–thiophene oligomers as n-channel semiconductors in organic field-effect and light-emitting transistors. *Chem. Mater.* **2014**, *26*, 6542–6556. [CrossRef]
4. de Zerio Mendaza, A.D.; Melianas, A.; Rossbauer, S.; Bäcke, O.; Nordstierna, L.; Erhart, P.; Olsson, E.; Anthopoulos, T.D.; Inganäs, O.; Müller, C. High-entropy mixtures of pristine fullerenes for solution-processed transistors and solar cells. *Adv. Mater.* **2015**, *27*, 7325–7331. [CrossRef]
5. Zhao, X.; Liu, T.; Cui, Y.; Hou, X.; Liu, Z.; Dai, X.; Kong, J.; Shi, W.; Dennis, T.J.S. Antisolvent-assisted controllable growth of fullerene single crystal microwires for organic field effect transistors and photodetectors. *Nanoscale* **2018**, *10*, 8170–8179. [CrossRef] [PubMed]
6. Zhang, F.; Hu, Y.; Schuettfort, T.; Di, C.-A.; Gao, X.; McNeill, C.R.; Thomsen, L.; Mannsfeld, S.C.B.; Yuan, W.; Sirringhaus, H.; et al. Critical role of alkyl chain branching of organic semiconductors in enabling solution-processed n-channel organic thin-film transistors with mobility of up to 3.50 cm^2 V^{-1} s^{-1}. *J. Am. Chem. Soc.* **2013**, *135*, 2338–2349. [CrossRef] [PubMed]
7. Welford, A.; Maniam, S.; Gann, E.; Thomsen, L.; Langford, S.J.; McNeill, C.R. Thionation of naphthalene diimide molecules: Thin-film microstructure and transistor performance. *Org. Electron.* **2018**, *53*, 287–295. [CrossRef]
8. Tilley, A.J.; Guo, C.; Miltenburg, M.B.; Schon, T.B.; Yan, H.; Li, Y.; Seferos, D.S. Thionation enhances the electron mobility of perylene diimide for high performance n-channel organic field effect transistors. *Adv. Funct. Mater.* **2015**, *25*, 3321–3329. [CrossRef]
9. Xu, X.; Yao, Y.; Shan, B.; Gu, X.; Liu, D.; Liu, J.; Xu, J.; Zhao, N.; Hu, W.; Miao, Q. Electron mobility exceeding 10 cm^2 V^{-1} s^{-1} and band-like charge transport in solution-processed n-channel organic thin-film transistors. *Adv. Mater.* **2016**, *28*, 5276–5283. [CrossRef] [PubMed]
10. Wu, Q.; Ren, S.; Wang, M.; Qiao, X.; Li, H.; Gao, X.; Yang, X.; Zhu, D. Alkyl chain orientations in dicyanomethylene-substituted 2,5-di(thiophen-2-yl)thieno-[3,2-b]thienoquinoid: Impact on solid-state and thin-film transistor performance. *Adv. Funct. Mater.* **2013**, *23*, 2277–2284. [CrossRef]
11. Facchetti, A.; Mushrush, M.; Katz, H.E.; Marks, T.J. n-Type building blocks for organic electronics: A homologous family of fluorocarbon-substituted thiophene oligomers with high carrier mobility. *Adv. Mater.* **2003**, *15*, 33–38. [CrossRef]
12. Facchetti, A.; Mushrush, M.; Yoon, M.-H.; Hutchison, G.R.; Ratner, M.A.; Marks, T.J. Building blocks for n-type molecular and polymeric electronics. Perfluoroalkyl-versus alkyl-functionalized oligothiophenes (nT; n = 2–6). Systematics of thin film microstructure, semiconductor performance, and modeling of majority charge injection in field-effect transistors. *J. Am. Chem. Soc.* **2004**, *126*, 13859–13874.

13. Yoon, M.-H.; Kim, C.; Facchetti, A.; Marks, T.J. Gate dielectric chemical structure–organic field-effect transistor performance correlations for electron, hole, and ambipolar organic semiconductors. *J. Am. Chem. Soc.* **2006**, *128*, 12851–12869. [CrossRef] [PubMed]
14. Capelli, R.; Toffanin, S.; Generali, G.; Usta, H.; Facchetti, A.; Muccini, M. Organic light-emitting transistors with an efficiency that outperforms the equivalent light-emitting diodes. *Nat. Mater.* **2010**, *9*, 496–503. [CrossRef] [PubMed]
15. Chang, J.-F.; Chen, W.-R.; Lai, Y.-C.; Chien, F.-C. Red phosphorescent trilayer organic light-emitting field-effect transistors with a wide recombination zone. *Jpn. J. Appl. Phys.* **2016**, *55*, 020304. [CrossRef]
16. Chang, J.-F.; Chen, W.-R.; Huang, S.-M.; Lai, Y.-C.; Lai, X.-Y.; Yang, Y.-W.; Wang, C.-H. High mobility ambipolar organic field-effect transistors with a nonplanar heterojunction structure. *Org. Electron.* **2015**, *27*, 84–91. [CrossRef]
17. Dholakia, G.R.; Meyyappan, M.; Facchetti, A.; Marks, T.J. Monolayer to multilayer nanostructural growth transition in n-type oligothiophenes on Au(111) and implications for organic field-effect transistor performance. *Nano Lett.* **2006**, *6*, 2447–2455. [CrossRef]
18. Facchetti, A.; Yoon, M.-H.; Stern, C.L.; Hutchison, G.R.; Ratner, M.A.; Marks, T.J. Building blocks for n-type molecular and polymeric electronics. Perfluoroalkyl- versus alkyl-functionalized oligothiophenes (nTs; n = 2–6). Systematic synthesis, spectroscopy, electrochemistry, and solid-state organization. *J. Am. Chem. Soc.* **2004**, *126*, 13480–13501. [CrossRef]
19. DeLongchamp, D.M.; Kline, R.J.; Lin, E.K.; Fischer, D.A.; Richter, R.J.; Lucas, L.A.; Heeney, M.; McCulloch, I.; Northrup, J.E. High carrier mobility polythiophene thin films: Structure determination by experiment and theory. *Adv. Mater.* **2007**, *19*, 833–837. [CrossRef]
20. Lue, J.-W.; Lin, Y.-H.; Yang, Y.-W. Growth and electronic structure studies of semiconducting thin films of fluorine-monosubstituted fused-thiophene derivative. *J. Electron Spectrosc. Relat. Phenom.* **2014**, *196*, 49–53. [CrossRef]
21. Dhez, O.; Ade, H.; Urquhart, S.G. Calibrated NEXAFS spectra of some common polymers. *J. Electron Spectrosc. Relat. Phenom.* **2003**, *128*, 85–96. [CrossRef]
22. Zharnikov, M.; Frey, S.; Heister, K.; Grunze, M. An extension of the mean free path approach to X-ray absorption spectroscopy. *J. Electron Spectrosc. Relat. Phenom.* **2002**, *124*, 15–24. [CrossRef]
23. Ohta, T.; Seki, K.; Yokoyama, T.; Morisada, I.; Edamatsu, K. Polarized XANES studies of oriented polyethylene and fluorinated polyethylenes. *Phys. Scr.* **1990**, *41*, 150–153. [CrossRef]
24. Väterlein, P.V.; Fink, R.; Umbach, E.; Wurth, W. Analysis of the x-ray absorption spectra of linear saturated hydrocarbons using the Xα scattered-wave method. *J. Chem. Phys.* **1998**, *108*, 3313–3320. [CrossRef]
25. Stöhr, J. *NEXAFS Spectroscopy*; Springer: Berlin, Germany, 1992.
26. Wang, C.-H.; Cheng, Y.-C.; Su, J.-W.; Fan, L.-J.; Huang, P.-Y.; Chen, M.-C.; Yang, Y.-W. Origin of high field-effect mobility in solvent-vapor annealed anthradithiophene derivative. *Org. Electron.* **2010**, *11*, 1947–1953. [CrossRef]
27. Brown, A.R.; Jarrett, C.P.; deLeeuw, D.M.; Matters, M. Field-effect transistors made from solution-processed organic semiconductors. *Synth. Met.* **1997**, *88*, 37–55. [CrossRef]
28. Street, R.A.; Northrup, J.E.; Salleo, A. Transport in polycrystalline polymer thin-film transistors. *Phys. Rev. B* **2005**, *71*, 165202. [CrossRef]
29. Salleo, A.; Chen, T.W.; Völkel, A.R.; Wu, Y.; Liu, P.; Ong, B.S.; Street, R.A. Intrinsic hole mobility and trapping in a regioregular poly(thiophene). *Phys. Rev. B* **2004**, *70*, 115311. [CrossRef]
30. Street, R.A. *Hydrogenated Amorphous Silicon*; Cambridge University Press: Cambridge, UK, 1991.
31. Chang, J.-F.; Sirringhaus, H.; Giles, M.; Heeney, M.; McCulloch, I. Relative importance of polaron activation and disorder on charge transport in high-mobility conjugated polymer field-effect transistors. *Phys. Rev. B* **2007**, *76*, 205204. [CrossRef]
32. Shi, J.; Wang, H.; Song, D.; Tian, H.; Geng, Y.; Yan, D. n-Channel, ambipolar, and p-channel organic heterojunction transistors fabricated with various film morphologies. *Adv. Funct. Mater.* **2007**, *17*, 397–400. [CrossRef]

© 2019 by the authors. Licensee MDPI, Basel, Switzerland. This article is an open access article distributed under the terms and conditions of the Creative Commons Attribution (CC BY) license (http://creativecommons.org/licenses/by/4.0/).

Commentary

Thin-Film Optical Devices Based on Transparent Conducting Oxides: Physical Mechanisms and Applications

Jiung Jang, Yeonsu Kang, Danyoung Cha, Junyoung Bae and Sungsik Lee *

Department of Electronics, Pusan National University, Pusan 46241, Korea; jee265@pusan.ac.kr (J.J.); yskang96@pusan.ac.kr (Y.K.); wdwn9277@pusan.ac.kr (D.C.); bjy0910@pusan.ac.kr (J.B.)
* Correspondence: sungsiklee@pusan.ac.kr; Tel.: +82-51-510-3123

Received: 25 February 2019; Accepted: 27 March 2019; Published: 3 April 2019

Abstract: This paper provides a review of optical devices based on a wide band-gap transparent conducting oxide (TCO) while discussing related physical mechanisms and potential applications. Intentionally using a light-induced metastability mechanism of oxygen defects in TCOs, it is allowed to detect even visible lights, eluding to a persistent photoconductivity (PPC) as an optical memory action. So, this PPC phenomenon is naturally useful for TCO-based optical memory applications, e.g., optical synaptic transistors, as well as photo-sensors along with an electrical controllability of a recovery speed with gate pulse or bias. Besides the role of TCO channel layer in thin-film transistor structure, a defective gate insulator can be another approach for a memory operation with assistance for gate bias and illuminations. In this respect, TCOs can be promising materials for a low-cost transparent optoelectronic application.

Keywords: transparent conducting oxides; oxygen defects; persistent photoconductivity; photo-sensors; optical synaptic devices

1. Introduction

Transparent conducting oxides (TCOs) are getting an intensive interest for a transparent display application since they exhibit high transparency arising from a wide band-gap [1–5]. Besides its high transparency, a low-temperature processability is another advantage suitable for display applications where a low melting-point substrate with a high flexibility is usually used [3,6,7]. This feature is mainly related to its isotropic bonding mechanism based on the overlap of s-orbitals for the conduction band, which is insensitive to bonding direction, providing high electron mobility. So, it can easily be connected even in a low-temperature fabrication. On the other hand, the conventional thin-film material for displays, e.g., amorphous silicon, has an anisotropic bonding nature due to its sp3 hybrid orbitals with a strong bonding directivity, thus difficult to be crystallized in a low temperature [6,8,9]. In this respect, TCOs have been considered as a strong candidate in futuristic electronics, including transparent and flexible displays [5,7,10–13].

However, TCOs have optical stability issues associated with oxygen defects, e.g., oxygen vacancies and interstitials even though these defects are playing the role as an electron donor to control the film conductivity during the deposition [8,14–16]. For example, the oxygen vacancy as an empty place of a bonding oxygen can be ionized under light with a relevant photon energy, giving free electrons [17–20]. And the deionization process, followed after removal of a light source, tends to be much slower than the ionization, suggesting a metastability, i.e., persistent photoconductivity (PPC) [21,22]. Moreover, this kind of metastable defects can also be newly created especially under a high energy illumination, e.g., ultra-violet (UV) [19,23]. From these circumstances, these oxygen defects are basically problematic, but simultaneously useful for photo-sensing applications if that slow recovery is overcome.

In this paper, we present a review of optical devices based on TCO thin films in terms of respective physical mechanisms and related applications. The key physical mechanism of TCO-based optical devices is associated with oxygen defects within TCO films (see Section 2). Since they are located within the band-gap, their location in energy is relevant to detect even visible light. This implies their optical ionization is a key absorption mechanism of photo-sensors. On the other hand, the deionization process after the removal of light, i.e., recovering process, is very slow in comparison with the ionization process, suggesting the PPC. This PPC is problematic for photo-sensors because it especially increases a dark signal level after the illumination, which can be overcome with applying a positive gate pulse [25]. As a positive aspect of the PPC, however, it is a memory action naturally induced under illumination. Indeed, optical memory devices, such as optical synaptic transistors, where a TCO (e.g., In-Ga-Zn-O, In-Zn-O, etc.) is a channel absorption layer, have been introduced using this phenomenon [26]. In another type of TCO-based optical synaptic thin-film transistors, a gate insulator, e.g., chitosan electrolyte or HfO_x, with mobile impurities or traps, is incorporated into a field-effect transistor structure along with a TCO channel layer, leading to a more distinguishable hysteresis for a strong memory action under light, depending on the polarity of gate bias [27,28]. In these respects, TCOs can be promising materials for a low-cost transparent optoelectronic application besides their basic usage for a transparent display application. These are discussed in Section 3.

2. Optical Properties and Instability Mechanism

2.1. Basic Optical Properties

TCOs have been considered as a strong candidate in futuristic electronics, including transparent and flexible displays, thanks to their high transparency (>80% in visible range) due to a large bandgap and a high mobility even under low temperature fabrication process associated with isotropic bonding nature of orbitals for conduction band bottom. However, a large carrier concentration, which is desired for a good conductivity, is thought to be an issue in optical transparency. Within the partially-filled conduction band, an intra-band transition occurs, thus an increase of optical absorption at long wavelengths [29–31]. Figure 1 shows a decrease in the transmission rate for IGZO films with a high carrier concentration. In contrast, at short wavelengths, this partially-filled conduction band gives rise to increasing the effective bandgap, as shown in Figure 2a, thus a blue-shifted optical absorption edge energy, which is called as a Burstein-Moss (BM) effect [24,32].

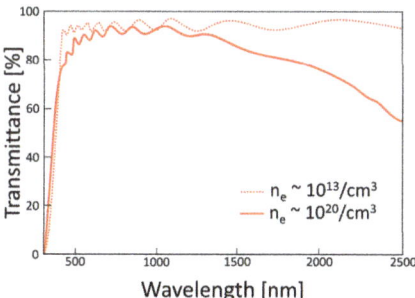

Figure 1. Optical transmission spectrum of the a-IGZO device for two different concentrations. A dashed line is $n_e \sim 10^{13}/cm^3$ and a solid line is $n_e \sim 10^{20}/cm^3$. Here, significant decrease of the transmittance of long wavelength light is found for the higher carrier concentration case (redrawn and adapted from [24]).

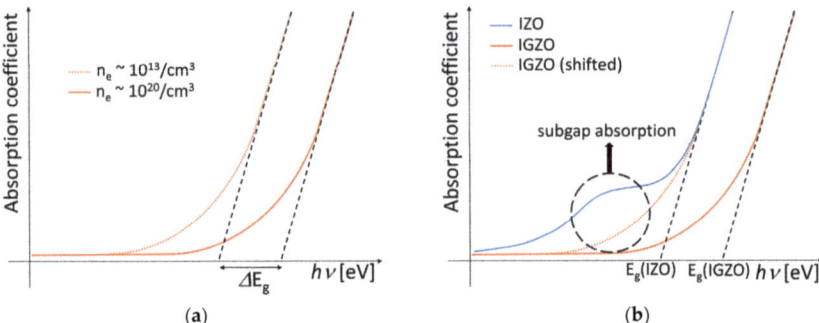

Figure 2. (a) Comparative Tauc plot for each carrier concentration ($n_e \sim 10^{13}/cm^3$ and $n_e \sim 10^{20}/cm^3$). (b) Conceptual Tauc plots to compare two different cases – large and small (IZO/IGZO) amount of oxygen defects (redrawn and adapted from [6]).

Different from these kinds of absorption related to carrier excitation, a subgap optical absorption also occurs due to ionization of subgap states [19–21,33,34]. Figure 2b depicts the conceptual Tauc plot for comparing amorphous IZO's and amorphous IGZO's subgap optical absorptions. The subgap absorption in TCOs is suggested to be mainly dependent on the presence of oxygen defects. As described in Figure 2b, the subgap absorption in IZO is higher compared to IGZO because the former contains relatively more oxygen defects [25,35]. The subgap absorption is not only problematic for transparency, it also causes interesting electrical issues, such as increase of carrier concentration, leading to an increase of the film conductivity, and a threshold voltage shift in the TFTs, electrical hysteresis, and the PPC. In this respect, we will further discuss this subgap absorption especially related to oxygen defects, and respective instability mechanisms under illumination in the following section.

2.2. Oxygen Defects and Optical Instability Mechanisms

Oxygen defects, such as oxygen vacancies (V_O) and oxygen interstitial (I_O), reside within deep subgap states, eluding to a sub-gap optical absorption even under a visible light (i.e., $h\nu < E_g$) while being ionized in the following processes, respectively, $V_O \rightarrow V_O^{2+} + 2e^-$ and $I_O^{2-} \rightarrow I_O^0 + 2e^-$ [21,30,36–39]. The excess electrons emitted during ionization process of oxygen defects contribute to the carrier concentration, increasing Fermi energy (E_F) and conductivity of the film. This affects negative threshold voltage shift in the operation of the TFTs, where TCOs are used for the channel layer, thus an increase of drain current [40–43]. Note that the density of oxygen defects can be controlled by changing the oxygen gas flow rate during the film deposition process [6,15,21]. With a sufficiently high oxygen flow rate, oxygen defects can be minimized whereas oxygen interstitials can newly be created with a very high oxygen flow rate, so that an optimization is needed. However, for an optical application, the presence of these defects is intentionally used as discussed here.

As illustrated in Figure 3a, interestingly, however, the ionization of oxygen defects involves relaxation of the lattice, suggesting a higher activation energy of the ionized defects compared to the neutral state defects [21,44,45]. It implies the ionized oxygen defects as metastable states. This metastability limits the recombination rate of the ionized defects, so the recovery process shows slower speed compared to the ionization. And this makes photo-enhanced conductivity persist for a relatively long time, depending on the temperature and bias condition, which is called PPC (see Figure 3b) [5,21,22,46]. This recovery trend is quantitatively analyzed by a stretched exponential function (SEF), as follows.

$$F(t) = A_{eff} \exp\left(-\left(\frac{t}{\tau_{eff}}\right)^\beta\right) \quad (1)$$

where A_{eff} is a weighting amplitude, τ_{eff} is an effective time constant, β is a stretched exponent whose range is from zero to unity. Note that A_{eff} is generally normalized as unity since it is related to the light intensity but not the photon energy. The PPC, which is a kind of optoelectronic hysteresis, is a memory action naturally induced with illumination. It implies a possibility for memory application without an additional capacitor structure. And the increase of film conductivity under illumination implies that TCO films can be used for a photo-sensor application as long as the PPC is eliminated to reset the reference signal level for a periodic sensing as a function of time. These operating concepts of devices are illustrated in Figure 4. And we discuss them in the following section.

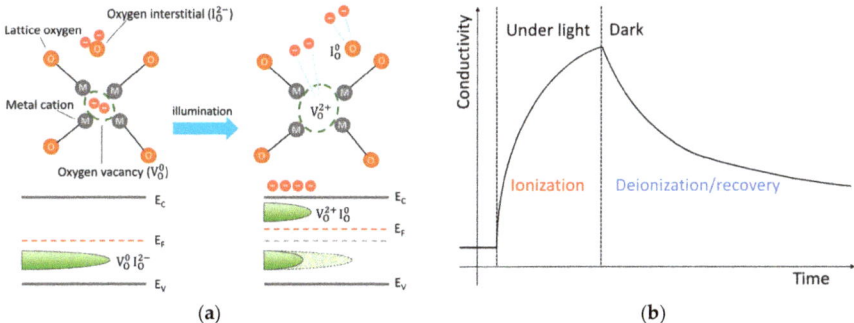

Figure 3. (a) Reaction diagram about ionizations of oxygen defects under illumination. Here, the lattice of the oxygen vacancy is depicted to be relaxed during the optically-induced ionization, thus an increase of the activation energy (redrawn and adapted from [21]). (b) Conductivity as a function of time. A seen, under illumination, conductivity of the transparent conducting oxide (TCO) film increases due to emitted free electrons during ionization process of the oxygen defects, and that under dark slowly decreases due to slow recombination of the free electrons.

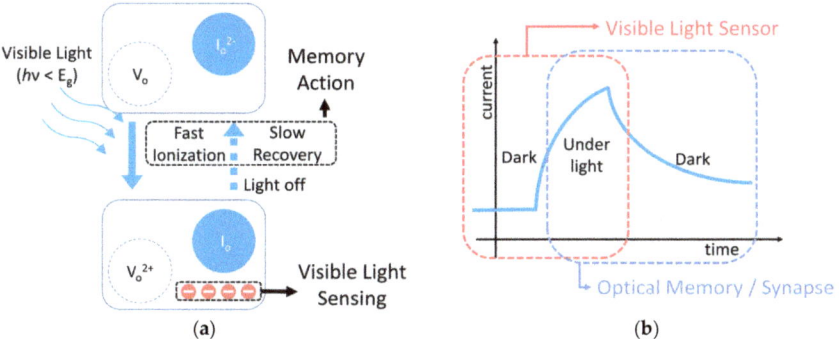

Figure 4. (a) Conceptual diagram to describe an optical memory-action and light-reactivity through the ionization of oxygen defects. (b) Indication of functional regions of the current as a function of time for each device.

3. Optoelectronic Applications

3.1. Photo-Sensors

In the previous section, it is shown that the TCO-based thin film can be applied for photo-sensors, detecting even visible lights which have a lower photon energy compared to the bandgap energy. With this feature, the TCO-based visible light photo-sensors can be integrated into transparent electronics. However, the PPC, associated with oxygen defects of the TCOs, increases the dark signal level after illumination (i.e., reference signal level), which can be problematic for a periodic photo-sensing, as mentioned in the previous section. In this section, we discuss the TCO-based photo-sensors with a possible solution for suppressing the PPC-related issue.

As an example seen in Figure 5, Sanghun Jeon et al. demonstrated a TCO-based device for photo-sensor arrays [25]. Here, an IZO channel layer, which has a relatively large amount of oxygen defects and respective high light-sensitivity, is employed for the main absorption layer. And two IGZO layers are adapted above and below the IZO layer to compensate a negative threshold voltage of the IZO layer-only TFT. So, it is found that the threshold voltage of IGZO/IZO/IGZO TFT is about 0 V. The authors set the device operating under a negative gate bias (V_{GS} = −7 V) and positive drain bias (V_{DS} = 10 V) for a distinguishable sensing level, raising I_{photo}/I_{dark} ratio, where I_{photo} is the photocurrent and I_{dark} is the dark current. When the device is illuminated, the oxygen defects in the channel layers are ionized, and simultaneously band-to-band excitation occurs, thus a free electron generation and raise of Fermi level. These cause negative shift of the threshold voltage, which is observable with an increase of the drain-source current. But the PPC is naturally occurred in the TCO films, and even reinforced by negative gate bias separating electrons and ionized oxygen defects, while increasing the dark signal level after removal of light. It can be problematic for a periodic light-sensing as above mentioned. To solve this problem, the positive pulse is applied to the gate terminal. It accumulates electrons at the channel interface, accelerating recombination of ionized oxygen defects with those induced electrons. As a result, the positive gate pulse leads to a fast recovery, making the dark signal level after illumination as same as that before illumination. Note that a positive gate pulsing scheme especially with a very high pulse height can lead to an adverse effect, such as charge trapping into the gate insulator, and more generally it can affect device reliability [15]. So, it should be carefully designed to avoid these side-effects.

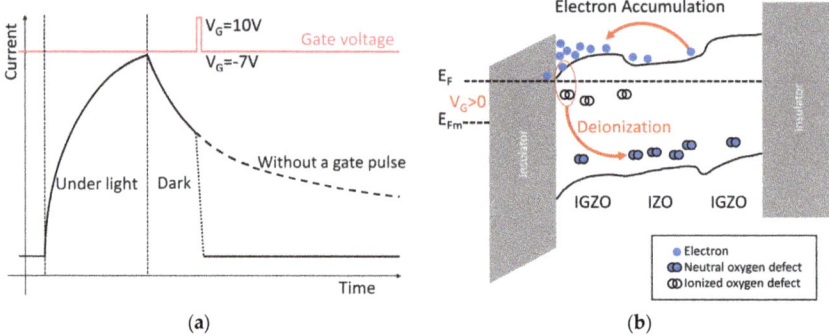

Figure 5. (a) Conceptual plot of the drain current of the TCO-based optical device. Here, the positive gate pulse makes dark signal level after illumination as same as the level before illumination as illustrated in this figure. (b) Energy band diagram of the device when positive gate pulse is applied. Here, it is described that the increased electron concentration of the film accelerates deionization of the oxygen defect (redrawn and adopted from [25]).

For another approach of the TCO-based thin film photo-sensor, there was an effort to improve an optical responsivity. Using the IZO-only TFT, where the IZO contains a relatively high concentration of oxygen defects, it is available to get a high optical responsivity and signal to noise ratio (SNR) [47]. Additionally, the blue light (460 nm) is intentionally chosen for the same reason in their measurements of transfer characteristics under illumination with different wavelengths.

3.2. Optical Synaptic Devices with an Optical Memory-Action

Though the PPC can be problematic for sensor applications, it is a memory action naturally induced under illumination. Inspired by this aspect, there have been several reports about synaptic devices mimicking a biological synapse. There are key functions that the synaptic devices should emulate, such as a spike-timing–dependent plasticity (STDP), short-term memory (STM) to long-term memory (LTM) transition, and facilitation (Figure 6a–d). Indeed, there are other possible types for synaptic devices (e.g., resistive switching devices [48,49], atomic-switching devices [50], and transistor-based devices [51–53]), which use an electrical signal, as the biological synapse does. Compared to these electrical-stimulus types, TCO-based synaptic devices, which use an optical stimulus, may offer much wider bandwidth, ultrafast signal transmission, low crosstalk, and being avoided from electrical shortcoming induced by a parasitic effect like the Miller effect [26,54]. Here, an issue to be resolved is still remaining to deal with the optical stimulus and current signal in a system level.

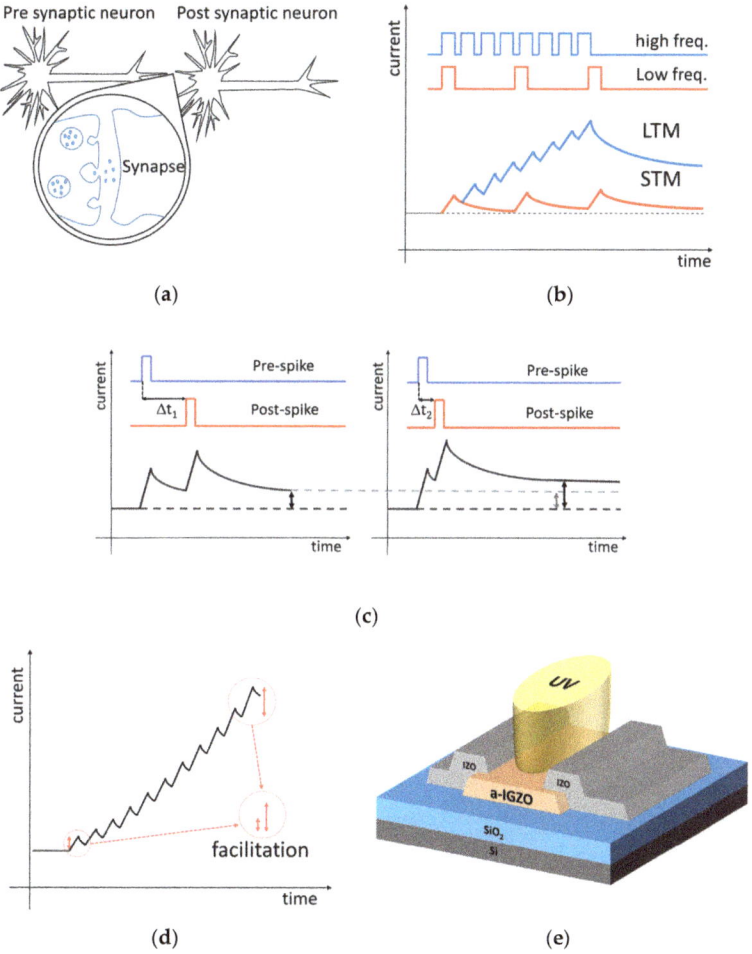

Figure 6. (**a**) Schematic showing neurons and synapses. (**b**) Frequency dependent short-term memory (STM) and long-term memory (LTM). (**c**) Plasticity variation with spike timing interval, spike-timing–dependent plasticity (STDP). (**d**) Facilitation under periodic pulses. (**e**) Device structure (redrawn and adopted from [26]).

A two-terminal synaptic device which uses the UV light pulses as the pre-synaptic stimulus has been demonstrated by Lee et al. [26]. Figure 6e shows its device structure. When the IGZO channel layer was exposed to the UV light, the drain current increased owing to the band-to-band excitation and ionization of oxygen defects at the same time. Once the UV light was turned off, current showed a slow decay due to the aforementioned PPC phenomenon. Here, it can be suggested that extra oxygen vacancies can also be created by the UV light since it is a high energy light [19,23], and these additionally contribute to the further increase and slow decay of the photocurrent. These slow decay suggests a memory action arising from the PPC, so that the device with the IGZO layer has an optical synaptic operation. In addition, the activation energy of an IGZO film increases under continuously illuminated UV pulses, incurring energy-state dependent facilitation (See Figure 6d). A slow decay of the post-illumination current and an increase of activation energy can lead to the frequency response of the photocurrent. This implies that it can also mimic timing-dependent memory actions of the brain synapse, such as the STDP and STM-LTM transition (see Figure 6b,c).

To electrically control PPC behavior for handling complex functions and distinguishable hysteresis, three-terminal devices were reported [27]. Here, the device adopted HfO_x as a gate insulator in TFT, which was thought to be a defective material. The hysteresis, which was induced by trapping and de-trapping of carriers at the IGZO/HfO_x interface and within the HfO_x layer [55], could assist the inherent PPC perform timing-related synaptic functions (Figure 6b–d)) in the IGZO layer. In addition, when a positive voltage pulse was applied to the gate electrode, the drain current had decreased due to electrons trapped by the defects. The experimental result of the effect is conceptually schemed in Figure 7c. Here, the authors demonstrated paired-pulse depression with this operation, which mimicked the decrease of synaptic weight under pre-synaptic stimulus in the synapse.

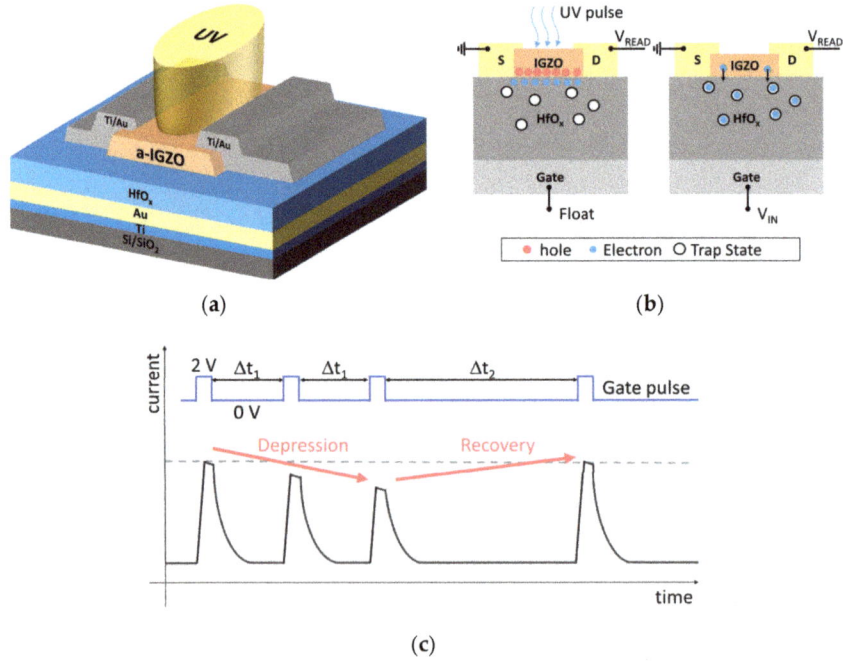

Figure 7. (a) Device structure. (b) Conceptual diagram of device operating under light pulse (**left**) and electrical pulse (**right**) applied. (c) Experimental result from the device operating under electrical pulses. With a short enough interval time between two pulses (Δt_1), drain current is decreased due to the trapped electrons in the HfO_x layer. On the other hand, with the long enough interval (Δt_2), the current level recovers from the depression effect (redrawn and adopted from [27]).

For another type of three-terminal device, the chitosan electrolyte is incorporated between the IGZO channel layer and ITO gate layer [28], as seen in Figure 8a. Under a positive gate bias, protons will accumulate near the interface between the IGZO and chitosan-electrolyte [56], and it helps the light-induced electrons stay at the interface even longer, reinforcing the PPC behavior (see Figure 8b). In this way, the amplitude of the optical response can be controlled with modulating the gate voltage. The experimental result of this operation is shown in Figure 8c. The authors call it a depression mode to potentiation mode transition. These examples discussed here clearly confirm that the TCO-based device with the PPC optical memory phenomena can be used as an optical synaptic transistor for an advanced neuromorphic system in a novel way, where an optoelectronic operating principle is employed rather than a just electrical operation.

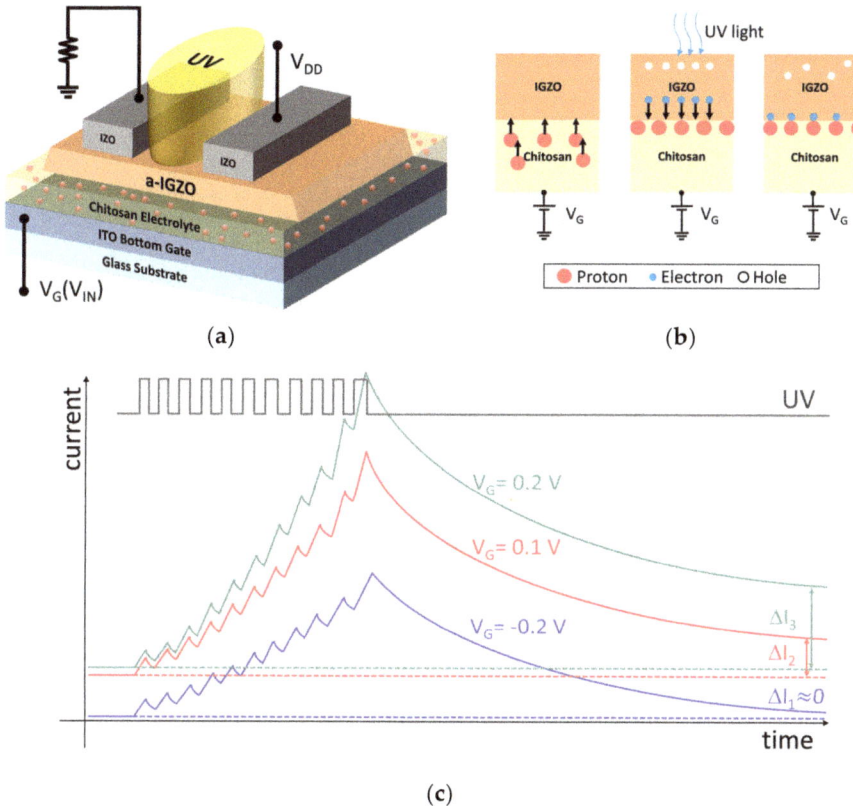

Figure 8. (a) Device structure and operating circuit. (b) Conceptual diagram of proton migration under gate bias and UV light illumination. (c) Experimental result from the device operating under different gate biases. The current differences ΔI_1, ΔI_2, and ΔI_3 mimic the variation of synaptic weight depending on the gate bias (redrawn and adopted from [28]).

4. Conclusions

We have discussed physical mechanisms and derived applications of the TCO-based thin-film devices in terms of how to intentionally use their optical properties associated with oxygen defects under illumination, such as the PPC. For photo sensing applications which can detect visible lights with TFT structures, it has been found that a positive voltage pulse is required to get rid of the PPC after illumination. In other words, a fast recombination of ionized oxygen defects, i.e., metastable states, is required. At the same time, it is also important to reinforce the persistency for an optical memory, such as artificial synapses with light stimuli. From applications referred in this work, it is expected that the TCO can be used as advanced multi-functional devices, while maintaining its advantages, such as a high transparency and high electrical performance.

Author Contributions: S.L. organized this work and finalized the whole manuscript. J.J. wrote the draft of the manuscript while discussing with other authors. J.J. and Y.K. prepared and reviewed Section 3, and D.C. and J.B. did Section 2.

Funding: This research was supported in part by Samsung Electronics (201900610001) and the Basic Science Research Program through the National Research Foundation of Korea (NRF) funded by the Ministry of Science, ICT & Future Planning (NRF-2018R1C1B6001688).

Conflicts of Interest: The authors declare no conflict of interest.

References

1. Nathan, A.; Lee, S.; Jeon, S.; Robertson, J. Amorphous Oxide Semiconductor TFTs for Displays and Imaging. *J. Disp. Technol.* **2014**, *10*, 917–927. [CrossRef]
2. Kamiya, T.; Hosono, H. Material characteristics and applications of transparent amorphous oxide semiconductors. *NPG Asia Mater.* **2010**, *2*, 15–22. [CrossRef]
3. Nomura, K.; Ohta, H.; Takagi, A.; Kamiya, T.; Hirano, M.; Hosono, H.J.N. Room-temperature fabrication of transparent flexible thin-film transistors using amorphous oxide semiconductors. *Nature* **2004**, *432*, 488. [CrossRef] [PubMed]
4. Hosono, H. 68.3: Invited Paper: Transparent Amorphous Oxide Semiconductors for High Performance TFT. In *Proceedings of SID Symposium Digest of Technical Papers*; Blackwell Publishing Ltd.: Oxford, UK, 2007; pp. 1830–1833.
5. Park, J.S.; Maeng, W.-J.; Kim, H.-S.; Park, J.-S. Review of recent developments in amorphous oxide semiconductor thin-film transistor devices. *Thin Solid Films* **2012**, *520*, 1679–1693. [CrossRef]
6. Lee, S.; Jeon, S.; Chaji, R.; Nathan, A. Transparent Semiconducting Oxide Technology for Touch Free Interactive Flexible Displays. *Proc. IEEE* **2015**, *103*, 644–664.
7. Kamiya, T.; Nomura, K.; Hosono, H. Present status of amorphous In-Ga-Zn-O thin-film transistors. *Sci. Technol. Adv. Mater.* **2010**, *11*, 044305. [CrossRef] [PubMed]
8. Kamiya, T.; Nomura, K.; Hosono, H. Origins of High Mobility and Low Operation Voltage of Amorphous Oxide TFTs: Electronic Structure, Electron Transport, Defects and Doping. *J. Disp. Technol.* **2009**, *5*, 468–483. [CrossRef]
9. Nathan, A.; Lee, S.; Jeon, S.; Song, I.; Chung, U.I.J.I.D. Transparent oxide semiconductors for advanced display applications. *Inf. Disp.* **2013**, *29*, 6–11. [CrossRef]
10. Yin, H.; Kim, S.; Kim, C.J.; Song, I.; Park, J.; Kim, S.; Park, Y. Fully transparent nonvolatile memory employing amorphous oxides as charge trap and transistor's channel layer. *Appl. Phys. Lett.* **2008**, *93*, 172109. [CrossRef]
11. Nomura, K.; Takagi, A.; Kamiya, T.; Ohta, H.; Hirano, M.; Hosono, H. Amorphous Oxide Semiconductors for High-Performance Flexible Thin-Film Transistors. *Jpn. J. Appl. Phys.* **2006**, *45*, 4303–4308. [CrossRef]
12. Fortunato, E.; Barquinha, P.; Martins, R. Oxide semiconductor thin-film transistors: A review of recent advances. *Adv. Mater.* **2012**, *24*, 2945–2986. [CrossRef]
13. Nathan, A.; Lee, S.; Jeon, S.; Song, I.; Chung, U.I. 3.1: Invited Paper: Amorphous Oxide TFTs: Progress and Issues. In *Proceedings of SID Symposium Digest of Technical Papers*; Blackwell Publishing Ltd.: Oxford, UK, 2012; pp. 1–4.
14. Janotti, A.; Van de Walle, C.G. Oxygen vacancies in ZnO. *Appl. Phys. Lett.* **2005**, *87*, 122102. [CrossRef]
15. Liu, L.; Mei, Z.; Tang, A.; Azarov, A.; Kuznetsov, A.; Xue, Q.-K.; Du, X. Oxygen vacancies: The origin ofn-type conductivity in ZnO. *Phys. Rev. B* **2016**, *93*, 235305. [CrossRef]
16. Leenheer, A.J.; Perkins, J.D.; van Hest, M.F.A.M.; Berry, J.J.; O'Hayre, R.P.; Ginley, D.S. General mobility and carrier concentration relationship in transparent amorphous indium zinc oxide films. *Phys. Rev. B* **2008**, *77*, 115215. [CrossRef]
17. Kim, S.; Kim, S.; Kim, C.; Park, J.; Song, I.; Jeon, S.; Ahn, S.-E.; Park, J.-S.; Jeong, J.K. The influence of visible light on the gate bias instability of In–Ga–Zn–O thin film transistors. *Solid-State Electron.* **2011**, *62*, 77–81. [CrossRef]

18. Jeon, J.-H.; Kim, J.; Ryu, M.-K. Instability of an Amorphous Indium Gallium Zinc Oxide TFT under Bias and Light Illumination. *J. Korean Phys. Soc.* **2011**, *58*, 158–162. [CrossRef]
19. Gurwitz, R.; Cohen, R.; Shalish, I. Interaction of light with the ZnO surface: Photon induced oxygen "breathing," oxygen vacancies, persistent photoconductivity, and persistent photovoltage. *J. Appl. Phys.* **2014**, *115*, 033701. [CrossRef]
20. Chowdhury, M.D.H.; Migliorato, P.; Jang, J. Light induced instabilities in amorphous indium–gallium–zinc–oxide thin-film transistors. *Appl. Phys. Lett.* **2010**, *97*, 173506. [CrossRef]
21. Lee, S.; Nathan, A.; Jeon, S.; Robertson, J. Oxygen Defect-Induced Metastability in Oxide Semiconductors Probed by Gate Pulse Spectroscopy. *Sci. Rep.* **2015**, *5*, 14902. [CrossRef]
22. Ghaffarzadeh, K.; Nathan, A.; Robertson, J.; Kim, S.; Jeon, S.; Kim, C.; Chung, U.I.; Lee, J.-H. Persistent photoconductivity in Hf–In–Zn–O thin film transistors. *Appl. Phys. Lett.* **2010**, *97*, 143510. [CrossRef]
23. Hensling, F.V.E.; Keeble, D.J.; Zhu, J.; Brose, S.; Xu, C.; Gunkel, F.; Danylyuk, S.; Nonnenmann, S.S.; Egger, W.; Dittmann, R. UV radiation enhanced oxygen vacancy formation caused by the PLD plasma plume. *Sci. Rep.* **2018**, *8*, 8846. [CrossRef]
24. Takagi, A.; Nomura, K.; Ohta, H.; Yanagi, H.; Kamiya, T.; Hirano, M.; Hosono, H. Carrier transport and electronic structure in amorphous oxide semiconductor, a-InGaZnO4. *Thin Solid Films* **2005**, *486*, 38–41. [CrossRef]
25. Jeon, S.; Ahn, S.E.; Song, I.; Kim, C.J.; Chung, U.I.; Lee, E.; Yoo, I.; Nathan, A.; Lee, S.; Robertson, J.; et al. Gated three-terminal device architecture to eliminate persistent photoconductivity in oxide semiconductor photosensor arrays. *Nat. Mater.* **2012**, *11*, 301–305. [CrossRef]
26. Lee, M.; Lee, W.; Choi, S.; Jo, J.W.; Kim, J.; Park, S.K.; Kim, Y.H. Brain-Inspired Photonic Neuromorphic Devices using Photodynamic Amorphous Oxide Semiconductors and their Persistent Photoconductivity. *Adv. Mater.* **2017**, *29*, 1700951. [CrossRef] [PubMed]
27. Wu, Q.; Wang, J.; Cao, J.; Lu, C.; Yang, G.; Shi, X.; Chuai, X.; Gong, Y.; Su, Y.; Zhao, Y.; et al. Photoelectric Plasticity in Oxide Thin Film Transistors with Tunable Synaptic Functions. *Adv. Electron. Mater.* **2018**, *4*, 1800556. [CrossRef]
28. Yang, Y.; He, Y.; Nie, S.; Shi, Y.; Wan, Q. Light Stimulated IGZO-Based Electric-Double-Layer Transistors For Photoelectric Neuromorphic Devices. *IEEE Electron Device Lett.* **2018**, *39*, 897–900. [CrossRef]
29. Facchetti, A.; Marks, T. *Transparent Electronics: From Synthesis to Applications*; John Wiley & Sons: Hoboken, NJ, USA, 2010.
30. Han, W.H.; Oh, Y.J.; Chang, K.J.; Park, J.-S. Electronic Structure of Oxygen Interstitial Defects in Amorphous In-Ga-Zn-O Semiconductors and Implications for Device Behavior. *Phys. Rev. Appl.* **2015**, *3*, 044008. [CrossRef]
31. Rhodes, C.; Franzen, S.; Maria, J.-P.; Losego, M.; Leonard, D.N.; Laughlin, B.; Duscher, G.; Weibel, S. Surface plasmon resonance in conducting metal oxides. *J. Appl. Phys.* **2006**, *100*, 054905. [CrossRef]
32. Yu, X.; Marks, T.J.; Facchetti, A. Metal oxides for optoelectronic applications. *Nat. Mater.* **2016**, *15*, 383–396. [CrossRef]
33. Jianke, Y.; Ningsheng, X.; Shaozhi, D.; Jun, C.; Juncong, S.; Shieh, H.D.; Po-Tsun, L.; Yi-Pai, H. Electrical and Photosensitive Characteristics of a-IGZO TFTs Related to Oxygen Vacancy. *IEEE Trans. Electron Devices* **2011**, *58*, 1121–1126. [CrossRef]
34. Nomura, K.; Kamiya, T.; Yanagi, H.; Ikenaga, E.; Yang, K.; Kobayashi, K.; Hirano, M.; Hosono, H. Subgap states in transparent amorphous oxide semiconductor, In–Ga–Zn–O, observed by bulk sensitive X-ray photoelectron spectroscopy. *Appl. Phys. Lett.* **2008**, *92*, 202117. [CrossRef]
35. Lee, S.; Jeon, S.; Robertson, J.; Nathan, A. How to achieve ultra high photoconductive gain for transparent oxide semiconductor image sensors. In Proceedings of the 2012 International Electron Devices Meeting, San Francisco, CA, USA, 10–13 December 2012.
36. Janotti, A.; Van de Walle, C.G. Fundamentals of zinc oxide as a semiconductor. *Rep. Prog. Phys.* **2009**, *72*, 126501. [CrossRef]
37. Jang, J.T.; Park, J.; Ahn, B.D.; Kim, D.M.; Choi, S.J.; Kim, H.S.; Kim, D.H. Study on the photoresponse of amorphous In-Ga-Zn-O and zinc oxynitride semiconductor devices by the extraction of sub-gap-state distribution and device simulation. *ACS Appl. Mater. Interfaces* **2015**, *7*, 15570–15577. [CrossRef]

38. Kamiya, T.; Nomura, K.; Hirano, M.; Hosono, H. Electronic structure of oxygen deficient amorphous oxide semiconductor a-InGaZnO4-x: Optical analyses and first-principle calculations. *Phys. Status Solidi (C)* **2008**, *5*, 3098–3100. [CrossRef]
39. Noh, H.-K.; Chang, K.J.; Ryu, B.; Lee, W.-J. Electronic structure of oxygen-vacancy defects in amorphous In-Ga-Zn-O semiconductors. *Phys. Rev. B* **2011**, *84*, 115205. [CrossRef]
40. Chong, E.-G.; Chun, Y.-S.; Kim, S.-H.; Lee, S.-Y. Effect of oxygen on the threshold voltage of a-IGZO TFT. *J. Electr. Eng. Technol.* **2011**, *6*, 539–542. [CrossRef]
41. Jeon, S.; Ahn, S.-E.; Song, I.; Jeon, Y.; Kim, Y.; Kim, S.; Choi, H.; Kim, H.; Lee, E.; Lee, S. Dual gate photo-thin film transistor with high photoconductive gain for high reliability, and low noise flat panel transparent imager. In Proceedings of the 2011 International Electron Devices Meeting, Washington, DC, USA, 5–7 December 2011.
42. Ahn, S.E.; Song, I.; Jeon, S.; Jeon, Y.W.; Kim, Y.; Kim, C.; Ryu, B.; Lee, J.H.; Nathan, A.; Lee, S.; et al. Metal oxide thin film phototransistor for remote touch interactive displays. *Adv. Mater.* **2012**, *24*, 2631–2636. [CrossRef] [PubMed]
43. Ghaffarzadeh, K.; Nathan, A.; Robertson, J.; Kim, S.; Jeon, S.; Kim, C.; Chung, U.-I.; Lee, J.H. Instability in threshold voltage and subthreshold behavior in Hf–In–Zn–O thin film transistors induced by bias-and light-stress. *Appl. Phys. Lett.* **2010**, *97*, 113504. [CrossRef]
44. Flewitt, A.J.; Powell, M.J. A thermalization energy analysis of the threshold voltage shift in amorphous indium gallium zinc oxide thin film transistors under simultaneous negative gate bias and illumination. *J. Appl. Phys.* **2014**, *115*. [CrossRef]
45. Jeon, S.; Song, I.; Lee, S.; Ryu, B.; Ahn, S.E.; Lee, E.; Kim, Y.; Nathan, A.; Robertson, J.; Chung, U.I. Origin of high photoconductive gain in fully transparent heterojunction nanocrystalline oxide image sensors and interconnects. *Adv. Mater.* **2014**, *26*, 7102–7109. [CrossRef] [PubMed]
46. Lee, S.; Nathan, A.; Robertson, J. Challenges in visible wavelength detection using optically transparent oxide semiconductors. In Proceedings of the SENSORS, 2012 IEEE, Taipei, Taiwan, 28–31 October 2012; pp. 1–4.
47. Liu, P.T.; Ruan, D.B.; Yeh, X.Y.; Chiu, Y.C.; Zheng, G.T.; Sze, S.M. Highly Responsive Blue Light Sensor with Amorphous Indium-Zinc-Oxide Thin-Film Transistor based Architecture. *Sci. Rep.* **2018**, *8*, 8153. [CrossRef] [PubMed]
48. Shi, J.; Ha, S.D.; Zhou, Y.; Schoofs, F.; Ramanathan, S. A correlated nickelate synaptic transistor. *Nat. Commun.* **2013**, *4*, 2676. [CrossRef] [PubMed]
49. Jiang, R.; Ma, P.; Han, Z.; Du, X. Habituation/Fatigue behavior of a synapse memristor based on IGZO-HfO2 thin film. *Sci. Rep.* **2017**, *7*, 9354. [CrossRef] [PubMed]
50. Ohno, T.; Hasegawa, T.; Tsuruoka, T.; Terabe, K.; Gimzewski, J.K.; Aono, M. Short-term plasticity and long-term potentiation mimicked in single inorganic synapses. *Nat. Mater.* **2011**, *10*, 591–595. [CrossRef] [PubMed]
51. Choi, H.-S.; Wee, D.-H.; Kim, H.; Kim, S.; Ryoo, K.-C.; Park, B.-G.; Kim, Y. 3-D Floating-Gate Synapse Array With Spike-Time-Dependent Plasticity. *IEEE Trans. Electron Devices* **2018**, *65*, 101–107. [CrossRef]
52. Zhu, L.Q.; Wan, C.J.; Guo, L.Q.; Shi, Y.; Wan, Q. Artificial synapse network on inorganic proton conductor for neuromorphic systems. *Nat. Commun.* **2014**, *5*, 3158. [CrossRef] [PubMed]
53. Gopalakrishnan, R.; Basu, A. Triplet Spike Time-Dependent Plasticity in a Floating-Gate Synapse. *IEEE Trans. Neural Netw. Learn. Syst.* **2017**, *28*, 778–790. [CrossRef]
54. Li, H.K.; Chen, T.P.; Liu, P.; Hu, S.G.; Liu, Y.; Zhang, Q.; Lee, P.S. A light-stimulated synaptic transistor with synaptic plasticity and memory functions based on InGaZnOx–Al2O3 thin film structure. *J. Appl. Phys.* **2016**, *119*, 244505. [CrossRef]
55. Zhu, W.J.; Ma, T.P.; Zafar, S.; Tamagawa, T. Charge trapping in ultrathin hafnium oxide. *IEEE Electron Device Lett.* **2002**, *23*, 597–599. [CrossRef]
56. Du, H.; Lin, X.; Xu, Z.; Chu, D. Electric double-layer transistors: A review of recent progress. *J. Mater. Sci.* **2015**, *50*, 5641–5673. [CrossRef]

© 2019 by the authors. Licensee MDPI, Basel, Switzerland. This article is an open access article distributed under the terms and conditions of the Creative Commons Attribution (CC BY) license (http://creativecommons.org/licenses/by/4.0/).

Article

Monotype Organic Dual Threshold Voltage Using Different OTFT Geometries

August Arnal [1], Carme Martínez-Domingo [1], Simon Ogier [2], Lluís Terés [1] and Eloi Ramon [1,*]

[1] Institut de Microelectrònica de Barcelona, IMB-CNM (CSIC), 08193 Bellaterra, Catalonia, Spain
[2] NeuDrive Limited, Biohub, Alderley Park, Macclesfield SK10 4TG, UK
* Correspondence: eloi.ramon@imb-cnm.csic.es

Received: 21 May 2019; Accepted: 27 June 2019; Published: 28 June 2019

Abstract: It is well known that organic thin film transistor (OTFT) parameters can be shifted depending on the geometry of the device. In this work, we present two different transistor geometries, interdigitated and Corbino, which provide differences in the key parameters of devices such as threshold voltage (V_T), although they share the same materials and fabrication procedure. Furthermore, it is proven that Corbino geometries are good candidates for saturation-mode current driven devices, as they provide higher I_{ON}/I_{OFF} ratios. By taking advantage of these differences, circuit design can be improved and the proposed geometries are, therefore, particularly suited for the implementation of logic gates. The results demonstrate a high gain and low hysteresis organic monotype inverter circuit with full swing voltage at the output.

Keywords: OTFT; interdigitated; Corbino; dual-threshold inverter; modelling; simulation

1. Introduction

Organic thin film transistors (OTFTs), employing organic dielectric and semiconductor layers, show considerable competitive advantages over their inorganic counterparts, such as low cost, low temperature processing and light-weight [1,2]. Compared with the conventional silicon dioxide-based devices, OTFTs with polymer dielectrics and blends of small-molecule semiconductors are ideally compatible with flexible substrates and solution processes. Apparently, solution-processable materials are very attractive because they are compatible with spin-coating, drop casting and printing technologies at room temperature and under ambient conditions. Meanwhile, this capability has practical advantages when coupled with photolithographic processes as a patterning techniques enable high-resolution, smaller process variations [3] and high-throughput [4]. By virtue of their excellent solution processability together with promising large-area coverage, OTFTs are attractive candidates in diverse applications [5–10], such as flexible displays [10,11] and radio frequency identification (RFID) [12], among others. Research on organic circuits has addressed in the last decades the development of inverters [13,14], logic gates, shift registers [14,15] and amplifiers [16–18]. In the majority of the works reporting organic circuits, the most widespread technology is single-threshold voltage (V_T) p-type only [19]. Although the most favourable route to obtain robust organic circuitry involves using organic complementary technology, this requires the matching of the n-type material in device performance which has proved difficult for complex circuits from a technological point of view [20]. Recently, the more widespread technology used to address the robustness of the organic circuits has been dual-V_T using only p-type organic semiconductors (OSC). Over recent years, several approaches have been elaborated in order to tune the electrical parameter V_T. For that purpose, two main routes have been considered: using chemical reactions at the dielectric/OSC interface or by modifying the OTFT geometry. For the former route, some examples include local chemical-doping of the channel [21], chemical modification of the channel interface through self-organizing polymer blending [22–24], and ultraviolet (UV) ozone and oxygen plasma treatments [25–27]. However, these chemical means present some drawbacks, such as the need for selective patterning of the reactive species/treatments or the fine control of the introduced charged states at the dielectric/OSC interface, which thus do not enable simple, large area processing. For the latter route related to OTFT geometry, good control over V_T was achieved by adding a double gate [28,29], enabling control and reversible shift over a wide range. However, this technology requires additional metal patterning by photolithography to vary the V_T through bottom and top electrodes substantially increasing the complexity of the devices and the circuits, as well as the number of layers and interconnections. Another approach to control V_T through the geometry is by means of the Corbino OTFT structure. Very few works have previously reported inorganic Corbino TFTs [30–33]. As OTFT based on small molecule OSC can present large diversity of crystal orientation and grain boundaries [34,35] and thus high V_T variability, Corbino OTFTs were developed to overcome this by having a circular channel making them more robust. In addition, Corbino shaped electrodes provide less overlaying areas in comparison with conventional interdigitated electrodes, which generates less parasitic capacitance, thus changing their electrical parameters compared with interdigitated electrodes. Other interesting features of Corbino OTFTs are the increment of the output resistance and the compatible fabrication with conventional interdigitated OTFTs without adding additional fabrication steps or chemical treatments. So far, Corbino TFTs have been uniquely employed as uniform current drivers in Active-Matrix Organic Light Emitting Diodes (AMOLED) to enhance the power efficiency [31,33].

Therefore, the ability to fabricate OTFTs using a fabrication methodology without involving chemical treatments with reproducible and uniform V_T is critical for practical circuit design. Improved circuit functionality is the main motivation for the present study, which considers both Corbino and interdigitated OTFT geometries as having dual-V_T behaviour. For the first time, dual-geometry allows innovative configuration of the logic gates presenting enhanced performance compared with single-geometry. Subsequently, dual-geometries were introduced and a substantial increase in gain and output voltage swing was achieved for inverters.

In addition, the efficient design of complex integrated circuits based on OTFTs requires preliminary characterization and modelling [36]. For this purpose, the development of accurate compact models is particularly appealing. In this paper, we first model and analyse the electrical characteristics and the performance parameters of the Corbino and interdigitated OTFTs possible in unipolar organic electronics using models based on the well-known Metal oxide field effect transistor (MOSFET) level 3.

2. Materials and Methods

The fabrication was carried out at the Centre for Process Innovation (CPI, Sedgefield, UK) using the Gen2 photolithography facilities. The Corbino and interdigitated OTFTs employ top gate bottom contact (TGBC) architecture in which the fabrication process begins with the spin-coating of the planarizing layer of a proprietary acrylate polymer (PCAF, from CPI) on carrier glass followed by a soft-bake, then a cross-linking process using UV light in a nitrogen environment, followed by hard-baking. Au (50 nm) were sputter coated and patterned using photolithography and wet etching to form drain and source electrodes. In order to increase the surface energy, the substrates were oxygen plasma-treated and then a self-assembled monolayer (SAM) of 3-fluoro-4-methoxythiophenol (Fluorochem, Glossop, United Kingdom) was deposited from a 10 mM solution in isopropanol and baked at 100 °C for 60 s. Afterwards, the OSC FlexOS^{TM} solution was spin-coated at 500 rpm for 10 s followed by 1000 rpm for 60 s and then a further bake at 100 °C for 60 s in order to obtain a 20 nm thick film. The OSC solution comprises a small molecule semiconductor and a high-k polymer semiconductor binder (Figure S1 in Supplementary Materials). The dielectric used in this work was Cytop CTL-809M (Asahi Glass, Tokyo, Japan) and it was diluted to 4.5% solids and spin-coated in order to obtain a 300 nm thick film and a gate capacitance of 6×10^{-9} F/cm^2. Au (50 nm) was thermally-evaporated, patterned using photolithography and wet etching. The gate-source/electrode overlapping area is 2.6×10^{-4} cm^2 and 1.4×10^{-5} cm^2 for Corbino and interdigitated devices, respectively. The unwanted areas of OSC and dielectric were patterned by oxygen reactive-ion etching plasma etching using the gate metal as a hard-mask. Subsequently, Polyvinyl alcohol (Sigma-Aldrich, Saint Quentin Fallavier, France) and SU-8 (MicroChem, Westborough, MA, United States of America) were deposited and patterned as a passivation layers. The metal interconnect layer (Au 50 nm) was sputtered and patterned to create electrical connections. Finally, the third protective passivation layer (SU-8, 450 nm thick) was deposited in a similar way to the first two passivation layers. All the layer thicknesses were measured through SEM images.

The devices were characterized with the requirements established for the model development, performing a wide range of values for the gate or drain voltage applying potential from −30 to 30 V. Using this procedure, experimental data on key parameters were successfully extracted. By fixing the polarization of the transistor with the same V_{GS} and V_{DS} the saturation region is assured, in our case −20 V, allowing implementation of the well-known extraction procedure for the saturation region of the transistors [37]. This procedure is based on plotting the square root of the drain current versus the gate voltage. In the polarization point of $V_{GS} = V_{DS}$, a straight line is plotted until the x-axis intercept where the threshold voltage value is extracted, see Figure S2B. The hole mobility was obtained dividing the slope value, in the polarization point of $V_{GS} = V_{DS}$, by the device total capacitance. Finally, the I_{ON}/I_{OFF} ratio was extracted from the ratio between the maximum current and the minimum current.

The dimensions of the Corbino devices were calculated as proposed in [38]. Defining R2 as the bigger radius and R1 the smaller one, the equivalent length of the device can be defined as:

$$L = (R2 - R1)$$

while the width of the transistor is determined with:

$$W = \frac{2 \cdot \pi}{\ln\left(\frac{R2}{R1}\right)} \cdot (R2 - R1)$$

The model implemented is based in MOSFET level 3 where some modifications had been implemented in order to fit the different transistors studied in this publication. Since the devices cannot be included in the under micrometre group, the effects of narrow width and short channel can be neglected, and the channel length modulation, λ, have to be considered, moreover the mobility degradation factor, θ, is included in terms of trapping in the channel. Finally, the bulk resistance has been implemented as, R_{BULK}, and added in all the stages of the compact model for keeping coherence between the different transistor behaviours. Additionally, the most common parameters such as the mobility, μ, the insulator capacitance, C_{INS}, the channel width, W, or the channel length, L, are included. The voltage and currents through the device are defined with the source, drain or gate terminals, having the difference between the first and the last defined terminals. The different regions and equations are introduced in the following equations:

For $V_{SG} < V_T$ in the cut-off region:

$$I_{SD} = R_{BULK} \cdot V_{SD}$$

$V_{SD} < V_{SG}-V_T$ for the linear region:

$$I_{SD} = \frac{\mu \cdot C_{INS} \cdot \frac{W}{L} \cdot \left(V_{SG} - V_t - \frac{V_{SD}}{2}\right) \cdot V_{SD} \cdot (1 + \lambda \cdot V_{SD})}{1 + \theta \cdot (V_{GS} - V_t)} + R_{BULK} \cdot V_{SD}$$

$V_{SD} >= V_{SG}-V_T$ when is in saturation region:

$$I_{SD} = \frac{\frac{1}{2} \cdot \mu \cdot C_{INS} \cdot \frac{W}{L} \cdot (V_{SG} - V_t)^2 \cdot (1 + \lambda \cdot V_{SD})}{1 + \theta \cdot (V_{GS} - V_t)} + R_{BULK} \cdot V_{SD}$$

The model has been extracted by using an in-house Matlab script that iterates with the equations and the experimental data of the transistors introducing a random stochastic value in a delimited range for an improved fitting in each device. These arbitrary values are initialized with the parameter extraction of the experimental data. Once the device is fitted with the MOSFET model, it is introduced in a Verilog—A module ready to be implemented in Virtuoso from Cadence (California, CA, United States of America).

All the electrical measurements were performed in ambient conditions and ambient light. The electrical characterization of the OTFTs was carried out using an Agilent B1500A Semiconductor Analyzer. The images were acquired using a light microscope DM4000 from Leica (Wetzlar, Germany).

3. Results

3.1. Interdigitated and Corbino OTFTs

For most of the basic research, the inverted/staggered and co-planar OTFT structures are the most commonly reported devices [39]. OTFTs can be implemented in different structures depending on the relative positions of the electrodes. Moreover, the electrodes in interdigitated configuration were predominantly adapted over the development of OTFT to account for the low conductivity of OSC. In this configuration, the source and drain electrodes are in a form of a comb, such that the finger of the drain comb is interdigitated with the fingers of the source comb providing larger current flow by increasing the channel width. Furthermore, it is worth mentioning that in the interdigitated geometry used in this work the gate layer totally overlies the source and drain electrodes. Figure 1Ai,Aii depicts the scheme and an optical image of the interdigitated electrodes fabricated in this work, respectively. The main advantage of this configuration is their efficiency concerning the ratio between the transistor surface and the transistor width [40]. However, such geometry is not entirely satisfactory in terms of performance since great efforts were made to enhance the alignment of OSC crystal perpendicularly along the Drain and Source (D/S) electrodes during the OSC deposition such as undergoing a nitrogen flow, temperature, and varying the solvent nature and ratio [41–43]. In order to overcome this limitation and for the sake of fabrication simplicity, OTFTs with circular channel, called Corbino OTFTs, are commonly proposed. The circular geometry experiences all orientations of the crystals providing less device variability. Although one intrinsic feature of Corbino geometry is a low gate-drain overlying area over their interdigitated geometry counterpart, in this work, the Corbino devices present an overlapped gate-source/drain capacitance of 2.6×10^{-4} cm^2 while the interdigitated devices present 1.4×10^{-5} cm^2. Figure 1Bi,Bii shows a cross-section and the optical image of interdigitated and Corbino geometries, respectively.

Interdigitated and Corbino OTFTs were electrically characterized with a W/L of 40 (W = 160 µm/L = 4 µm) and 34 (W = 680 µm/L = 20 µm), respectively. For a better understanding of the device performances according to the geometry, the transfer curves were compared by normalizing the drain current with the W/L ratio, called channel dimension (CD). Since the devices were fabricated simultaneously on the same substrate, they show the same layer thicknesses and characteristics, thus, the observed differences in the transfer-output electrical characteristics are solely attributed to OTFT geometry.

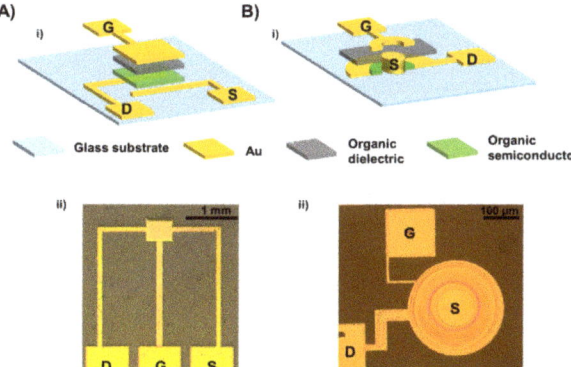

Figure 1. (**A**) (i) 3D scheme of the Interdigitated geometry and (ii) its optical image; (**B**) (i) 3D scheme of the Corbino geometry and (ii) its optical image. D, G and S refer to drain, gate and source electrodes, respectively.

The measured and simulated transfer and the output characteristics of the interdigitated and Corbino OTFTs are shown in Figure 2. As can be observed in Figure 2A, the model correctly predicts with high precision the transfer curves for both geometries. In addition, the simulated output curves fit well with the measurement by providing a smooth linear-to-saturation transition, as shown in Figure 2B. Thus, the used model can be widely applied to OTFTs with interdigitated and Corbino geometries serving as a practical and versatile tool for organic-based circuit development. Figure 2C,D show the measured output electrical characteristics for both geometries. The maximum drain current, named as I_{ON}, is significantly larger for Corbino devices than interdigitated devices, as can be observed in the zoom-in in Figure 2D, presenting values of 4.8 μA and 1.4 μA, respectively, for $V_{DS} = -20$ V and $V_{GS} = -20$ V. The same tendency is corroborated in the linear regime as can be seen in the transfer curve in Figure S3, where the higher current is attributed to the Corbino geometry. Despite the fact that effective carrier mobility is affected by the dimensions of the channel length, the Corbino OTFTs tends slightly to saturate. In contrast, the output drain current of interdigitated OTFTs showed a linear increase with V_{DS} and did not saturate. The unsaturated output characteristics in interdigitated devices were also clearly of larger channel length (see Figure S4 Supporting Information), thus, this phenomenon was not totally governed by short-channel effects. Indeed, the output characteristics of both geometries exhibited an obvious I_{DS} offset at $V_{GS} = 0$ V indicating an appreciable conduction, more pronounced in interdigitated OTFTs, due to a parasitic conduction path from the drain to the source. Due to this fact, the depletion mode current, which is called I_{OFF}, can be achieved by applying a positive voltage such that the conduction is suppressed. The depletion mode current in interdigitated geometry was reached by applying a $V_{GS} = -20$ V while in Corbino geometry the voltage applied was reduced to $V_{GS} = -10$ V, as can be compared in Figure 2C and the inset of Figure 2D. The I_{OFF} displayed are about one order of magnitude larger for interdigitated compared with Corbino devices, for $V_{GS} = 0$ V and $V_{DS} = -20$ V. In comparison with the Corbino devices reported previously [32], the circular devices developed in this work yielded similar output curves regardless the bias configuration, i.e., the inner-drain or outer-drain condition (see Figure S5, Supporting Information).

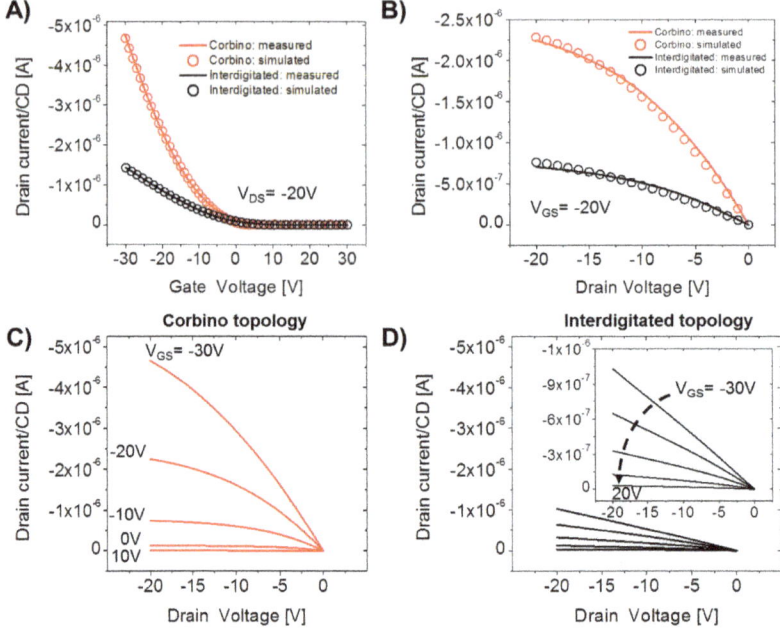

Figure 2. Experimental transfer (**A**) and output (**B**) characteristics for Corbino and interdigitated geometries normalized to Channel Dimension (CD) for $V_{DS} = -20$ V and $V_{GS} = -20$ V, respectively. The circular symbol refers to the theoretical transfer and output characteristics simulated by using a MOSFET level 3. (**C**) Experimental output characteristics of Corbino geometry devices. (**D**) Experimental output characteristics of interdigitated geometry devices where gate voltage (V_{GS}) were swept by -10 V from -30 V to 20 V. The output graph for interdigitated geometry is rescaled in the inset graph for better comparison. The channel length is 4 μm and 20 μm for interdigitated and the Corbino devices, respectively. Both have almost the same W/L ratio.

In order to gain insight into the electrical performance of both geometries, effective mobility, V_T, and I_{ON}/I_{OFF} current ratio were investigated as a figure of merit of interdigitated and Corbino geometries as shown in Figure 3A–C, respectively. Notably, there is a big difference in the effective hole mobility between the devices with different geometries. This result is due to the Corbino geometry despite presenting higher channel length compared with interdigitated OTFTs. The extracted effective hole mobility for Corbino geometries is about 1.61 cm²/V·s, and for interdigitated devices is about 0.71 cm²/V·s. Clearly, the OTFTs with Corbino geometry performed V_T closer to 0 V in the 4–8 V range, whereas interdigitated geometry performed V_T ranging from 8 to 14 V. Despite showing different threshold voltage for each geometry, the turn on voltage is the same for both devices since they share the same materials and fabrication procedure. Turn-on voltage has been extracted from the maximum point of the derivative in the logarithmic scale of the transfer curve and this result is included in Figure S6. Corbino devices present less variability of V_{ON} values than the interdigitated geometry devices. Regarding this difference, both types have very similar values for the turn on voltage, which is linked to the materials of the devices. Both devices work in accumulation mode, a result in good agreement with the improvement of the effective hole mobility in Corbino devices, which makes the V_T nearer zero. Interestingly, in the Corbino OTFTs, the V_T and effective hole mobility possess uniformity in terms of electrical parameter dispersion, which indicates that circular channel geometry is more robust in delivering good performance over large areas despite the polycrystalline nature of the small-molecules. A large number of methods have been developed to reduce the OTFT's current variability induced by random grain orientations of the organic semiconductor [44–46]. The semiconductor used in this work, FlexOS^{TM}, presents large crystal sizes in the range of tens of micrometres. As the grain structures have different grain orientations, the extreme grain-to-channel alignments were solved by a circular-shaped channel providing higher effective mobility and thus, lower V_T. The channel length is in the range of the grain domains: L = 4 µm and 20 µm for interdigitated and Corbino geometries, respectively. In order to measure the uniformity of the devices, the relative standard deviation (RSD), defined as the ratio of the standard deviation to the mean drain current, was used. In particular, our results show that the overall RSD in drain current for interdigitated OTFT and Corbino OTFT were found to be 65.18% and 12.4%, respectively. Apart from the reduction in variability, an enhanced effective mobility and threshold voltage were achieved for Corbino OTFTs. Other electrical parameters of the devices have been studied as can be seen in Figure 3C, where the I_{ON}/I_{OFF} ratio is degraded for interdigitated devices owing to the presence of parasitic current in the *OFF* state. The I_{ON}/I_{OFF} ratios for Corbino geometries are about (10^7), two orders of magnitude higher than those of interdigitated devices. For logic gates, the I_{ON}/I_{OFF} ratio remains an important parameter that must be taken into account. Thus, a high current modulation ratio is a more important requirement than the high mobility for programmable electronic circuits [43,47–50]. This behaviour is supported by the fact that Corbino geometry eliminates parasitic sources to drain current flows, hence, allowing an enhanced I_{ON}/I_{OFF} ratio. This phenomenon, previously reported in [33], is explained by the increment of the differential output conductance of Corbino devices, which is defined as $\delta I_{DS}/\delta V_{DS}$ and can be described as follows: For circular OTFTs, the W/L relationship remains fairly constant because the channel length modulation is compensated by the channel width modulation. Hence, beyond the pinch-off, the differential output resistance of the interdigitated TFT is finite; whereas, that of the Corbino TFT is infinite, which is in agreement with previous works [31,33] and with the results obtained in Figure 3D.

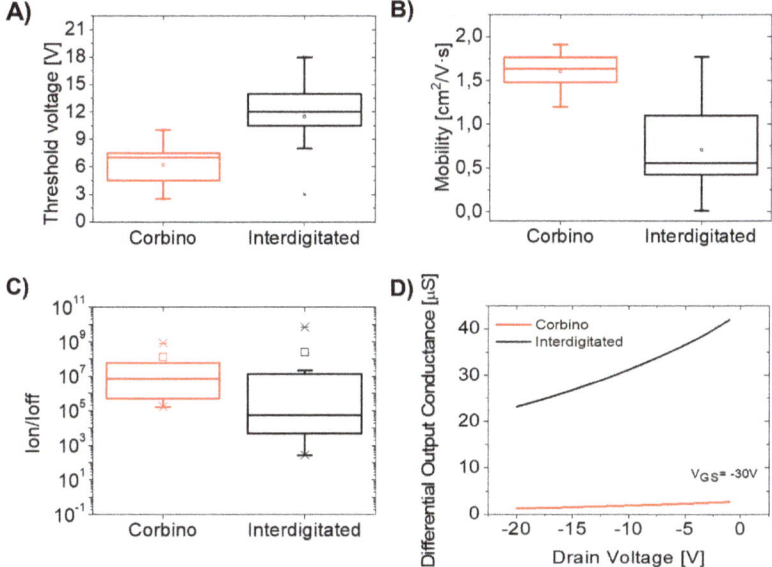

Figure 3. Electrical parameters such as (**A**) threshold voltage, (**B**) effective hole mobility and (**C**) I_{ON}/I_{OFF} ratio for Corbino and interdigitated geometries. The central mark represents the median, box limits indicate the 25th and 75th percentiles, and whiskers extend to the 5th and 9th percentiles. (**D**) Differential output conductance as a function of drain voltage at $V_{GS} = -30$ V.

To conclude, these results reflect the fact that the electrical performances actually correlate with the overall geometric parameter ($\mu_h C_{INS} W/L$) as supported by the higher the effective hole mobility (μ_h), the higher the V_T and, the higher the I_{ON}/I_{OFF} ratios for Corbino geometry. For these reasons, and as a novelty compared with widely-reported logic gates based on interdigitated OTFTs, an inverter logic gate was implemented in further studies in this work.

3.2. Unipolar Organic Dual-Geometry Threshold Voltage Inverter

The Inverter is considered to be the most basic logic circuit element for Complementary Metal-Oxide Semiconductor (CMOS) technology [51]. However, unipolar logic circuits are widely employed in the organic electronics field.

Unipolar based circuits which present either n-type or p-type OSC, require a pull-up (n) or pull-down (p) load transistor polarized in *ON* state and, an input-controlled drive transistor which logically inverts its input [52,53]. Therefore, for a low voltage (in terms of Boolean algebra, it is known as '0') the input produces a high voltage (in terms of Boolean algebra, it is known as '1') at the output, and vice versa [54]. In order to obtain a pull-down load transistor different geometries can be implemented. One of the most common geometries is based on the short-circuit of the gate and drain terminals yielding a fixed gate-to-source voltage, i.e., $V_{GS} = 0$ V, which turns the load transistor into a diode. In fact, this configuration is referred to as diode-load or depletion-mode load configuration [55,56]. Based on the results of several works [57] which demonstrated that higher gain can be achieved by diode-load topology, this work will consider a diode-load topology for the inverter.

Table 1. Comparison of different topologies of OTFT design styles for basic inverters with key parameters.

Inverter Topologies	CMOS	Pseudo-CMOS	Dual-Gate	Diode-Load	Dual-Threshold
Transistors	2	4	2	2	2
Power rails	2	3	2	2	2
Noise Margin	Most Robust	High Robust	Robust	Poor Robust	Medium Robust
Voltage swing	Full Swing	Full Swing	Non-Full Swing	Non-Full Swing	Almost-Full Swing
Power	Dynamic	Static and Dynamic	Static and Dynamic	Static and Dynamic	Static and Dynamic
Device Type	Complementary type	Mono-type	Mono-type	Mono-type	Mono-type

Different inverter topologies have been presented in the literature for monotype systems as shown in Table 1. From the simplest one using only one transistor and a resistor, to a more complex system where multiple devices are required. The most employed inverter topology for CMOS [58] technology involves two complementary transistors, meaning n-channel and p-channel. The advantage of this topology is full swing in the voltage achieved due to the alternative switching between *ON* and *OFF* state of the load and driver. Furthermore, this configuration allows a high gain with a significant robustness to the noise margin. The power consumption reaches minimum values because of the negligible power consumption in the static state of the inverter but, as a drawback, the fabrication processes involved are complex. For the monotype devices, the pseudo-CMOS [57,59] approach is the most robust topology since the polarization of the last transistor stage allows them to achieve full swing at the output. Despite this benefit, this topology requires four transistors and more power rails increasing the power consumption, which is detrimental for energy efficiency. In order to reduce the number of transistors and power sources required, the diode-load [60–63] is used because it operates with two unipolar transistors. However, this topology is less robust and presents non-full swing voltage at the output. Using dual-gate devices, [29,64,65] the diode-load configuration can have improved noise margins since the drive transistor is more robust against noise. The drawback of dual-gate transistors is the requirement of a more complex fabrication process and the inclusion of an additional power rail in the design.

The solution presented in this work is the use of a dual-threshold configuration [66,67] which provides a lower noise margin and higher voltage swing at the output compared to the diode-load and the dual-gate topology, respectively. Dual-threshold technology has been studied for several years and applied to the silicon technology, achieving power consumption reductions [68,69] and increasing the performance of the system in different aspects as delay [68,70], robust designs [71], asynchronous circuits [72], speed improvement [73] or glitch minimization [74]. The interdigitated and Corbino devices developed in this work present different electrical characteristics, such as V_T. Taking advantage of this behaviour, a dual-threshold configuration was implemented using these two geometries in order to obtain enhanced performance and reduced power rails. An interesting feature of this approach is the extreme simplicity for fabricating logic gates because the electrical parameters of the transistors are totally dependent of its geometry, allowing a unique manufacturing process for the entire circuit area.

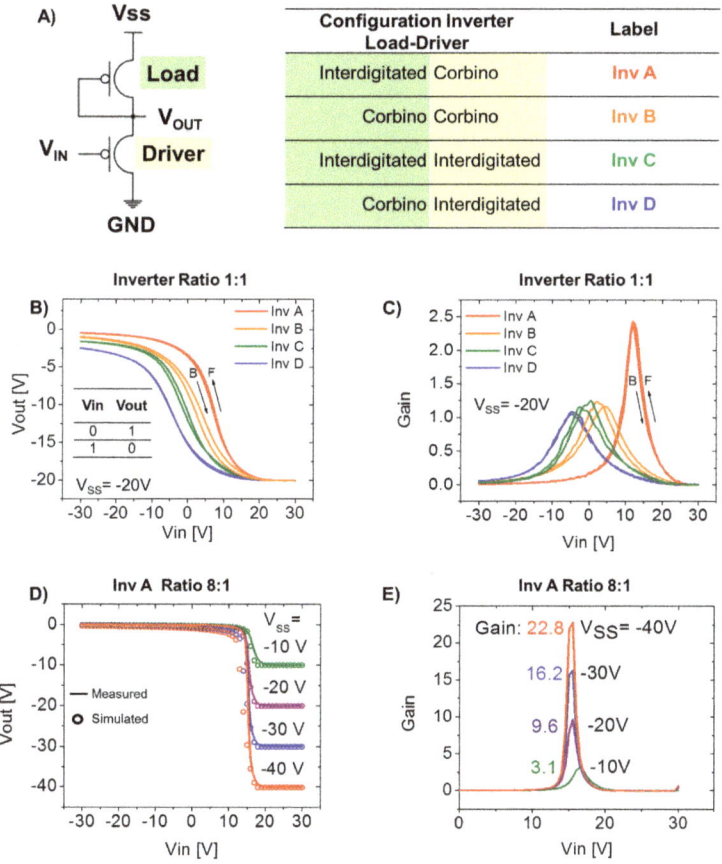

Figure 4. (A) Inverter circuits by using different combinations of transistor geometries: (B) Voltage transfer characteristics (VTC); (C) and gain. (D) VTC for Inv A geometry with a ratio 8:1; and (E) the respective gain. "F" and "B" denote forward and backward sweeps, respectively.

Figure 4A shows the schematic of the inverter. The load transistor is biased to the *ON* state, which allows the current to flow through itself by providing low resistance and the driver transistor switches between *ON* and *OFF* state controlled by the input. A figure of merit of the inverter circuit is the gain, which is defined as the maximum slope of the transfer curve of the system. Regarding gain, four different combinations of the two geometries for the driver and load were configured and characterized. The dimensions ratio, W/L, of the devices were equal (ratio 1:1) to be sure that the results were not influenced by the dimensions, as can be seen in Figure 4B. It can be noted that the configurations with a single type of geometry, i.e., interdigitated-interdigitated or Corbino-Corbino, provide higher hysteresis than the others. The best combination obtained was placing the interdigitated geometry as a load and the Corbino geometry as the driver (called Inv A), where the gain was double the rest of circuit configurations, as shown in Figure 4C. Moreover, this combination geometry offered the full output voltage swing. Taking into account the driver transistor commutating between the *ON* and *OFF* states, the Corbino geometry as used as a driver because it provides higher performance than the interdigitated geometry, in terms of the I_{ON}/I_{OFF} ratio, allowing high current in the *ON* state and a negligible current in the *OFF* mode. Furthermore, the interdigitated transistor working as a load provided lower output resistance and, thus, a lower voltage drop at the output of the system allowing it to reach ground (GND) and V_{SS} values. Additionally, both geometries presented different threshold voltages, which induced a more resistive load device when the driver was in an *ON* state. These behaviours occurred owing to the lower V_{SD} and resistance in the *OFF* state of the driver, permitting a full swing of the output voltage thus reducing the hysteresis of the inverter.

To assess the maximum gain obtained by the topology of Inv A, a variation on the ratio of dimensions between the load transistor and the drive transistor was implemented by increasing the driver 8 times with respect to the load, as can be seen in Figure 4D,E. The study of the electrical behaviour of the inverter for four different V_{SS} revealed a link between the power supply and gain. In fact, the increased gain correlated with a higher power supply and at the same time delivered higher swing and higher slope. Moreover, the simulation of electrical behaviour for Inv A depending on the V_{SS} was implemented in Virtuoso and by introducing the generated models of each geometry, as can be observed in Figure 4D. Well-fitting measurements at the transitions were obtained for low V_{SS}, meanwhile the model correctly predicted the V_{SS} and GND values for all V_{SS}. The increment in the ratio between the driver and the load provided almost four times higher gain compared with the previous 1:1 ratio. This configuration presented a gain of 22.8 of the V_{SS} to −40 V, as can be observed in Figure 4E.

4. Discussion

The implementation of inverter circuits in the organic electronics field has been a hot topic in recent years and different configurations have been tested using diverse circuit topologies. Until now, there have been no publications on improved inverter behaviour using two different geometries of transistors. However, as shown in this work improved inverter behaviour was achieved by exploiting the different electrical parameters, i.e., I_{ON}/I_{OFF} ratio, effective mobility and threshold voltage of interdigitated and Corbino geometries. In this work the two OTFTs were fabricated by means of the same fabrication process and materials to produce a novel, monotype dual threshold voltage organic inverter with two different transistor geometries for the load and the driver. An experimental comparison of geometries was presented, in which the same dimensions for the driver and inverter load were used in order to corroborate the advantages of the dual threshold technology. The best configuration with maximum gain and lowest hysteresis in the structure was found to be with a Corbino geometry as the driver and an interdigitated geometry as the load. This configuration provided a gain of up to 22.8 with a power supply of −40 V. The achieved gain was higher than most of the inverters presented with monotype systems. Additionally, the transistors were modelled with a modified MOSFET level 3 to improve fitting with the experimental data. Finally, the models were employed to simulate a circuit of the inverter and a good fit was achieved with the experimental data.

Supplementary Materials: The following are available online at http://www.mdpi.com/2073-4352/9/7/333/s1, Figure S1: Molecule structures of the small molecule organic semiconductor and the high-k polymer semiconductor binder used for the blend. Figure S2: Square root and logarithmic scale transfer curves of both geometries types for the two behavior of the transistors: (a) Lineal region, and (b) saturation region including the threshold voltage extraction procedure representation. Figure S3: Experimental transfer characteristics for Corbino and interdigitated geometries normalized to channel dimension (CD) for V_{DS} = −2 V. Figure S4: Experimental output characteristics of interdigitated geometry devices where gate voltage (V_{GS}) were swept by −10 V from −30 V to 10 V. The channel length is 10 µm. Figure S5: Output characteristics of the Corbino TFT for the two drain-bias conditions: (a) The inner circle as source and outer ring as drain; and (b) the inner circle as drain and outer ring as source. Figure S6: Turn on voltage of both geometries from the curve transfer curve with V_{DS} = −20 V. The central mark represents the median, box limits indicate the 25th and 75th percentiles, and whiskers extend to 5th and 9th percentiles.

Author Contributions: Device characterization, model development, circuit simulation and writing—original draft preparation: A.A.; Methodology, formal analysis and writing—review and editing: C.M.-D.; OTFT manufacturing: S.O.; funding acquisition and project administration: L.T. Conceptualization and supervision: E.R.

Funding: This research was funded by MICIU-Spain, grant number RTI2018–102070-B-C21.

Acknowledgments: The authors would like to thank Neudrive, Ltd. (UK) for providing the OTFT samples used on this work.

Conflicts of Interest: The authors declare no conflict of interest.

References

1. Tang, W.; Huang, Y.; Han, L.; Liu, R.; Su, Y.; Guo, X.; Yan, F. Recent progress in printable organic field effect transistors. *J. Mater. Chem. C* **2019**, *7*, 790–808. [CrossRef]
2. Sirringhaus, H. 25th anniversary article: Organic field-effect transistors: The path beyond amorphous silicon. *Adv. Mater.* **2014**, *26*, 1319–1335. [CrossRef] [PubMed]
3. Chang, J.S.; Facchetti, A.F.; Reuss, R.; Member, S.; Facchetti, A.F.; Reuss, R. A Circuits and Systems Perspective of Organic/Printed Electronics: Review, Challenges, and Contemporary and Emerging Design Approaches. *IEEE J. Emerg. Sel. Top. Circuits Syst.* **2017**, *7*, 7–26. [CrossRef]
4. Gamota, D.R. Printed organic and molecular electronics. *Mater. Today* **2004**, *7*, 53.
5. Ling, M.M.; Bao, Z. Thin film deposition, patterning, and printing in organic thin film transistors. *Chem. Mater.* **2004**, *16*, 4824–4840. [CrossRef]
6. Chabinyc, M.L.; Salleo, A. Materials requirements and fabrication of active matrix arrays of organic thin-film transistors for displays. *Chem. Mater.* **2004**, *16*, 4509–4521. [CrossRef]
7. Chou, W.Y.; Kuo, C.W.; Cheng, H.L.; Chen, Y.R.; Tang, F.C.; Yang, F.Y.; Shu, D.Y.; Liao, C.C. Effect of surface free energy in gate dielectric in pentacene thin-film transistors. *Appl. Phys. Lett.* **2006**, *89*, 2004–2007. [CrossRef]
8. Forrest, S.R. The path to ubiquitous and low-cost organic electronic appliances on plastic. *Nature* **2014**, *428*, 911–918. [CrossRef]
9. Anthopoulos, T.D.; De Leeuw, D.M.; Cantatore, E.; Setayesh, S.; Meijer, E.J.; Tanase, C.; Hummelen, J.C.; Blom, P.W.M. Organic complementary-like inverters employing methanofullerene-based ambipolar field-effect transistors. *Appl. Phys. Lett.* **2004**, *85*, 4205–4207. [CrossRef]
10. Meijer, E.J.; De Leeuw, D.M.; Setayesh, S.; Van Veenendaal, E.; Huisman, B.H.; Blom, P.W.M.; Hummelen, J.C.; Scherf, U.; Klapwijk, T.M. Solution-processed ambipolar organic field-effect transistors and inverters. *Nat. Mater.* **2003**, *2*, 678–682. [CrossRef]
11. Yoneya, N.; Ono, H.; Ishii, Y.; Himori, K.; Hirai, N.; Abe, H.; Yumoto, A.; Kobayashi, N.; Nomoto, K. Flexible electrophoretic display driven by solution-processed organic thin-film transistors. *J. Soc. Inf. Disp.* **2012**, *20*, 143. [CrossRef]
12. Myny, K.; Steudel, S.; Vicca, P.; Beenhakkers, M.J.; van Aerle, N.A.J.M.; Gelinck, G.H.; Genoe, J.; Dehaene, W.; Heremans, P. Plastic circuits and tags for 13.56 MHz radio-frequency communication. *Solid. State. Electron.* **2009**, *53*, 1220–1226. [CrossRef]
13. Cao, Q.; Kim, H.S.; Pimparkar, N.; Kulkarni, J.P.; Wang, C.; Shim, M.; Roy, K.; Alam, M.A.; Rogers, J.A. Medium-scale carbon nanotube thin-film integrated circuits on flexible plastic substrates. *Nature* **2008**, *454*, 495–500. [CrossRef]

14. Fluxman, S.M. Design and performance of digital polysilicon thin-film-transistor circuits on glass. *IEE Proc. Circuits, Devices Syst.* **1994**, *141*, 56. [CrossRef]
15. Bae, B.S.; Choi, J.W.; Oh, J.H.; Jang, J. Level shifter embedded in drive circuits with amorphous silicon TFTs. *IEEE Trans. Electron Devices* **2006**, *53*, 494–498.
16. Greening, B.; Kuo, C.C.; Sheraw, C.D.; Gundlach, D.J.; Cuomo, F.P.; Klauk, H.; Campi, J.; Nichols, J.A.; Huang, J.R.; Jia, L.; et al. Analog and digital circuits using organic thin-film transistors on polyester substrates. *IEEE Electron Device Lett.* **2000**, *21*, 534–536.
17. Fuketa, H.; Yoshioka, K.; Shinozuka, Y.; Ishida, K.; Yokota, T.; Matsuhisa, N.; Inoue, Y.; Sekino, M.; Sekitani, T.; Takamiya, M.; et al. 1 µm-Thickness Ultra-Flexible and High Electrode-Density Surface Electromyogram Measurement Sheet With 2 V Organic Transistors for Prosthetic Hand Control. *IEEE Trans. Biomed. Circuits Syst.* **2014**, *8*, 824–833. [CrossRef]
18. Ishida, K.; Huang, T.C.; Honda, K.; Sekitani, T.; Nakajima, H.; Maeda, H.; Takamiya, M.; Someya, T.; Sakurai, T. A 100-V AC energy meter integrating 20-V organic CMOS digital and analog circuits with a floating gate for process variation compensation and a 100-V organic pMOS rectifier. *IEEE J. Solid-State Circuits* **2012**, *47*, 301–309. [CrossRef]
19. Feng, L.; Tang, W.; Zhao, J.; Cui, Q.; Jiang, C.; Guo, X. All-solution-processed low-voltage organic thin-film transistor inverter on plastic substrate. *IEEE Trans. Electron Devices* **2014**, *61*, 1175–1180. [CrossRef]
20. Myny, K.; Beenhakkers, M.; van Aerle, N.; Gelinck, G.; Genoe, J.; Dehaene, W.; Heremans, P. Unipolar organic transistor circuits made robust by dual-gate technology. *IEEE J. Solid-State Circuits* **2011**, *46*, 1223–1230. [CrossRef]
21. Kergoat, L.; Herlogsson, L.; Piro, B.; Pham, M.C.; Horowitz, G.; Crispin, X.; Berggren, M. Tuning the threshold voltage in electrolyte-gated organic field-effect transistors. *Proc. Natl. Acad. Sci. USA* **2012**, *109*, 8394–8399. [CrossRef] [PubMed]
22. Hoon Kim, S.; Rim Hwang, H.; Joong Kwon, H.; Jang, J. Unipolar depletion-load organic circuits on flexible substrate by self-organized polymer blending with 6, 13-bis (triisopropylsilylethynyl) pentacene using ink-jet printing. *Appl. Phys. Lett.* **2012**, *100*, 053302. [CrossRef]
23. Pernstich, K.P.; Haas, S.; Oberhoff, D.; Goldmann, C.; Gundlach, D.J.; Batlogg, B.; Rashid, A.N.; Schitter, G. Threshold voltage shift in organic field effect transistors by dipole monolayers on the gate insulator. *J. Appl. Phys.* **2004**, *96*, 6431–6438. [CrossRef]
24. Kobayashi, S.; Nishikawa, T.; Takenobu, T.; Mori, S.; Shimoda, T.; Mitani, T.; Shimotani, H.; Yoshimoto, N.; Ogawa, S.; Iwasa, Y. Control of carrier density by self-assembled monolayers in organic field-effect transistors. *Nat. Mater.* **2004**, *3*, 317–322. [CrossRef] [PubMed]
25. Choi, J.M.; Im, S. Optimum channel thickness of rubrene thin-film transistors. *Appl. Phys. Lett.* **2008**, *93*, 2006–2009. [CrossRef]
26. Wang, A.; Kymissis, I.; Bulović, V.; Akinwande, A.I. Tunable threshold voltage and flatband voltage in pentacene field effect transistors. *Appl. Phys. Lett.* **2006**, *89*, 112109. [CrossRef]
27. Wang, A.; Kymissis, I.; Bulović, V.; Akinwande, A.I. Engineering density of semiconductor-dielectric interface states to modulate threshold voltage in OFETs. *IEEE Trans. Electron Devices* **2006**, *53*, 9–13. [CrossRef]
28. Takamiya, M.; Sekitani, T.; Kato, Y.; Kawaguchi, H.; Someya, T.; Sakurai, T. An Organic FET SRAM for Braille Sheet Display with Back Gate to Increase Static Noise Margin. *2006 IEEE Int. Solid State Circuits Conf.-Dig. Tech. Pap.* **2006**, *42*, 1060–1069.
29. Koo, J.B.; Ku, C.H.; Lim, J.W.; Kim, S.H. Novel organic inverters with dual-gate pentacene thin-film transistor. *Org. Electron.* **2007**, *8*, 552–558. [CrossRef]
30. Ogier, S.D.; Matsui, H.; Feng, L.; Simms, M.; Mashayekhi, M.; Carrabina, J.; Terés, L.; Tokito, S. Uniform, high performance, solution processed organic thin-film transistors integrated in 1 MHz frequency ring oscillators. *Org. Electron.* **2018**, *54*, 40–47. [CrossRef]
31. Mativenga, M.; Choe, Y.; Um, J.; Jang, J. *Corbino Oxide TFTs for Flexible AMOLED Display Stability*; Wiley Publishing: San Francisco, CA, USA, 2016; pp. 1147–1150.
32. Lee, H.; Yoo, J.S.; Kim, C.D.; Chung, I.J.; Kanicki, J. Asymmetric electrical properties of Corbino a-Si:H TFT and concepts of its application to flat panel displays. *IEEE Trans. Electron Devices* **2007**, *54*, 654–662. [CrossRef]

33. Mativenga, M.; Ha, S.H.; Geng, D.; Kang, D.H.; Mruthyunjaya, R.K.; Heiler, G.N.; Tredwell, T.J.; Jang, J. Infinite output resistance of corbino thin-film transistors with an amorphous-InGaZnO active layer for large-area AMOLED displays. *IEEE Trans. Electron Devices* **2014**, *61*, 3199–3205. [CrossRef]
34. Chen, J.; Tee, C.K.; Shtein, M.; Anthony, J.; Martin, D.C. Grain-boundary-limited charge transport in solution-processed 6,13 bis (tri-isopropylsilylethynyl) pentacene thin film transistors. *J. Appl. Phys.* **2008**, *103*, 114513. [CrossRef]
35. Steiner, F.; Poelking, C.; Niedzialek, D.; Andrienko, D.; Nelson, J. Influence of orientation mismatch on charge transport across grain boundaries in tri-isopropylsilylethynyl (TIPS) pentacene thin films. *Phys. Chem. Chem. Phys.* **2017**, *19*, 10854–10862. [CrossRef] [PubMed]
36. Fayez, M.; Morsy, K.; Sabry, M. Simulation of Organic Thin Film Transistor at both Device and Circuit Levels. *Int. Conf. Aerosp. Sci. Aviat. Technol.* **2019**, *16*, 1–6. [CrossRef]
37. Dobrescu, L.; Petrov, M.; Dobrescu, D.; Ravariu, C. Threshold voltage extraction methods for MOS transistors. In Proceedings of the 2000 International Semiconductor Conference, Sinaia, Romania, 10–14 October 2000; pp. 371–374.
38. Munteanu, D.; Cristoloveanu, S.; Hovel, H. Circular Pseudo-Metal Oxide Semiconductor Field Effect Transistor in Silicon-on-Insulator Analytical Model, Simulation, and Measurements. *Electrochem. Solid-State Lett.* **1999**, *2*, 242–243. [CrossRef]
39. Guo, X.; Xu, Y.; Ogier, S.; Ng, T.N.; Caironi, M.; Perinot, A.; Li, L.; Zhao, J.; Tang, W.; Sporea, R.A.; et al. Current Status and Opportunities of Organic Thin-Film Transistor Technologies. *IEEE Trans. Electron Devices* **2017**, *64*, 1906–1921. [CrossRef]
40. Mashayekhi, M. Inkjet-Configurable Gate Arrays. Towards Application Specific Printed Electronic Circuits. Available online: https://www.tesisenred.net/bitstream/handle/10803/402272/moma1de1.pdf?sequence=1&isAllowed=y (accessed on 27 June 2019).
41. Fujisaki, Y.; Takahashi, D.; Nakajima, Y.; Nakata, M.; Tsuji, H.; Yamamoto, T. Alignment Control of Patterned Organic Semiconductor Crystals in Short-Channel Transistor Using Unidirectional Solvent Evaporation Process. *IEEE Trans. Electron Devices* **2015**, *62*, 2306–2312. [CrossRef]
42. Bi, S.; He, Z.; Chen, J.; Li, D. Solution-grown small-molecule organic semiconductor with enhanced crystal alignment and areal coverage for organic thin film transistors. *AIP Adv.* **2015**, *5*, 077170. [CrossRef]
43. Tisserant, J.N.; Wicht, G.; Göbel, O.F.; Bocek, E.; Bona, G.L.; Geiger, T.; Hany, R.; Mezzenga, R.; Partel, S.; Schmid, P.; et al. Growth and alignment of thin film organic single crystals from dewetting patterns. *ACS Nano* **2013**, *7*, 5506–5513. [CrossRef]
44. Goto, O.; Tomiya, S.; Murakami, Y.; Shinozaki, A.; Toda, A.; Kasahara, J.; Hobara, D. Organic single-crystal arrays from solution-phase growth using micropattern with nucleation control region. *Adv. Mater.* **2012**, *24*, 1117–1122. [CrossRef] [PubMed]
45. Liu, Y.; Zhao, X.; Cai, B.; Pei, T.; Tong, Y.; Tang, Q.; Liu, Y. Controllable fabrication of oriented micro/nanowire arrays of dibenzo-tetrathiafulvalene by a multiple drop-casting method. *Nanoscale* **2014**, *6*, 1323–1328. [CrossRef] [PubMed]
46. Li, Y.; Ji, D.; Liu, J.; Yao, Y.; Fu, X.; Zhu, W.; Xu, C.; Dong, H.; Li, J.; Hu, W. Quick Fabrication of Large-area Organic Semiconductor Single Crystal Arrays with a Rapid Annealing Self-Solution-Shearing Method. *Sci. Rep.* **2015**, *5*, 1–9. [CrossRef] [PubMed]
47. Sou, A.; Jung, S.; Gili, E.; Pecunia, V.; Joimel, J.; Fichet, G.; Sirringhaus, H. Programmable logic circuits for functional integrated smart plastic systems. *Org. Electron.* **2014**, *15*, 3111–3119. [CrossRef]
48. Xu, W.; Liu, Z.; Zhao, J.; Xu, W.; Gu, W.; Zhang, X.; Qian, L.; Cui, Z. Flexible logic circuits based on top-gate thin film transistors with printed semiconductor carbon nanotubes and top electrodes. *Nanoscale* **2014**, *6*, 14891–14897. [CrossRef] [PubMed]
49. Sirringhaus, H.; Friend, R.H.; Li, X.C.; Moratti, S.C.; Holmes, A.B.; Feeder, N. Bis(dithienothiophene) organic field-effect transistors with a high ON/OFF ratio. *Appl. Phys. Lett.* **1997**, *71*, 3871–3873. [CrossRef]
50. Chandar Shekar, B.; Lee, J.; Rhee, S.W. Organic thin film transistors: Materials, processes and devices. *Korean J. Chem. Eng.* **2007**, *21*, 267–285. [CrossRef]
51. Briseno, A.L.; Tseng, R.J.; Li, S.H.; Chu, C.W.; Yang, Y.; Falcao, E.H.L.; Wudl, F.; Ling, M.M.; Chen, H.Z.; Bao, Z.; et al. Organic single-crystal complementary inverter. *Appl. Phys. Lett.* **2006**, *89*, 1–4. [CrossRef]

52. Wu, Q.; Zhang, J.Y.; Qin, R.Q. Design Considerations for Digital Circuits Using Organic Thin Film Transistors on a Flexible Substrate. In Proceedings of the 2006 IEEE International Symposium on Circuits and Systems, Island of Kos, Greece, 21–24 May 2006; Volume 1, pp. 1267–1270.
53. Lugli, P.; Csaba, G.; Erlen, C. Modeling of circuits and architectures for molecular electronics. *J. Comput. Electron.* **2009**, *8*, 410–426. [CrossRef]
54. Krumm, J. CIRCUIT ANALYSIS METHODOLOGY FOR ORGANIC TRANSISTORS. Available online: https://opus4.kobv.de/opus4-fau/files/627/diss.pdf (accessed on 27 June 2019).
55. Lee, C.A.; Jin, S.H.; Jung, K.D.; Lee, J.D.; Park, B.G. Full-swing pentacene organic inverter with enhancement-mode driver and depletion-mode load. *Solid. State. Electron.* **2006**, *50*, 1216–1218. [CrossRef]
56. Brown, A.R.; De Leeuw, D.M.; Havinga, E.E.; Pomp, A. Field-effect transistors made from solution-processed organic semiconductors. *Synth. Met.* **1994**, *68*, 65–70. [CrossRef]
57. Huang, T.C.; Fukuda, K.; Lo, C.M.; Yeh, Y.H.; Sekitani, T.; Someya, T.; Cheng, K.T. Pseudo-CMOS: A design style for low-cost and robust flexible electronics. *IEEE Trans. Electron Devices* **2011**, *58*, 141–150. [CrossRef]
58. Zeng, W.J.; Zhou, X.Y.; Pan, X.J.; Song, C.L.; Zhang, H.L. High performance CMOS-like inverter based on an ambipolar organic semiconductor and low cost metals. *AIP Adv.* **2013**, *3*, 012101. [CrossRef]
59. Fukuda, K.; Sekitani, T.; Yokota, T.; Kuribara, K.; Huang, T.C.; Sakurai, T.; Zschieschang, U.; Klauk, H.; Ikeda, M.; Kuwabara, H.; et al. Organic pseudo-CMOS circuits for low-voltage large-gain high-speed operation. *IEEE Electron Device Lett.* **2011**, *32*, 1448–1450. [CrossRef]
60. Kim, S.H.; Jang, J.; Jeon, H.; Yun, W.M.; Nam, S.; Park, C.E. Hysteresis-free pentacene field-effect transistors and inverters containing poly (4-vinyl phenol-co-methyl methacrylate) gate dielectrics. *Appl. Phys. Lett.* **2008**, *92*, 1–4. [CrossRef]
61. Kim, S.H.; Choi, D.; Chung, D.S.; Yang, C.; Jang, J.; Park, C.E.; Park, S.H.K. High-performance solution-processed triisopropylsilylethynyl pentacene transistors and inverters fabricated by using the selective self-organization technique. *Appl. Phys. Lett.* **2008**, *93*, 1–4. [CrossRef]
62. Koo, J.B.; Ku, C.H.; Lim, S.C.; Kim, S.H.; Lee, J.H. Hysteresis and threshold voltage shift of pentacene thin-film transistors and inverters with Al_2O_3 gate dielectric. *Appl. Phys. Lett.* **2007**, *90*, 1–4. [CrossRef]
63. Park, S.K.; Member, S.; Anthony, J.E.; Jackson, T.N. Solution-Processed TIPS-Pentacene Organic Thin-Film-Transistor Circuits. *IEEE Electron Device Lett.* **2007**, *28*, 2007–2009. [CrossRef]
64. Myny, K.; Beenhakkers, M.; van Aerle, N.; Gelinck, G.; Genoe, J.; Dehaene, W.; Heremans, P. Robust digital design in organic electronics by dual-gate technology. *IEEE Int. Solid-State Circuits Conf. Dig. Tech. Pap.* **2010**, *8*, 140–141.
65. Spijkman, M.J.; Myny, K.; Smits, E.C.P.; Heremans, P.; Blom, P.W.M.; De Leeuw, D.M. Dual-gate thin-film transistors, integrated circuits and sensors. *Adv. Mater.* **2011**, *23*, 3231–3242. [CrossRef]
66. Nausieda, I.; Ryu, K.K.; He, D.D.; Akinwande, A.I.; Bulovic, V.; Sodini, C.G. Mixed-signal organic integrated circuits in a fully photolithographic dual threshold voltage technology. *IEEE Trans. Electron Devices* **2011**, *58*, 865–873. [CrossRef]
67. Nausieda, I.; Ryu, K.K.; He, D.D.; Akinwande, A.I.; Bulović, V.; Sodini, C.G. Dual threshold voltage organic thin-film transistor technology. *IEEE Trans. Electron Devices* **2010**, *57*, 3027–3032. [CrossRef]
68. Wei, L.; Chen, Z.; Johnson, M.; Roy, K.; De, V. Design and optimization of low voltage high performance dual threshold CMOS circuits. In Proceedings of the 35th annual Design Automation Conference, San Francisco, CA, USA, 15–19 June 1998; pp. 489–494.
69. Sundararajan, V.; Parhi, K.K. Low power synthesis of dual threshold voltage CMOS VLSI circuits. In Proceedings of the 1999 International Symposium on Low Power Electronics and Design, San Diego, CA, USA, 17 August 1999; pp. 139–144.
70. Ghosh, A.; Ghosh, D. Optimization of static power, leakage power and delay of full adder circuit using dual threshold MOSFET based design and T-spice simulation. In Proceedings of the 2009 International Conference on Advances in Recent Technologies in Communication and Computing, Kerala, India, 27–28 October 2009; pp. 903–905.
71. Islam, A.; Akram, M.W.; Pable, S.D.; Hasan, M. Design and analysis of robust dual threshold CMOS full adder circuit in 32nm technology. In Proceedings of the 2010 International Conference on Advances in Recent Technologies in Communication and Computing, Kottayam, India, 16–17 October 2010; pp. 418–420.
72. Ghavami, B.; Pedram, H. Design of dual threshold voltages asynchronous circuits. In Proceedings of the 2008 international symposium on Low Power Electronics & Design, Bangalore, India, 11–13 August 2008; p. 185.

73. Wang, C.; Huang, C.; Lee, C. A Low Power High-Speed 8-Bit Pipelining CLA Design Using Dual-Threshold Voltage Domino Logic. *IEEE Trans. Very Large Scale Integr. VLSI Syst.* **2008**, *16*, 594–598. [CrossRef]
74. Slimani, M.; Matherat, P.; Mathieu, Y. A dual threshold voltage technique for glitch minimization. In Proceedings of the 2012 19th IEEE International Conference on Electronics, Circuits, and Systems (ICECS 2012), Seville, Spain, 9–12 December 2012; pp. 444–447.

© 2019 by the authors. Licensee MDPI, Basel, Switzerland. This article is an open access article distributed under the terms and conditions of the Creative Commons Attribution (CC BY) license (http://creativecommons.org/licenses/by/4.0/).

Editorial

Thin Film Transistor

Ray-Hua Horng

Institute of Electronics, National Chiao Tung University, Hsinchu 30010, Taiwan; rhh@nctu.edu.tw

Received: 25 July 2019; Accepted: 9 August 2019; Published: 9 August 2019

Abstract: The special issue is "Thin Film Transistor". There are eight contributed papers. They focus on organic thin film transistors, fluorinated oligothiophenes transistors, surface treated or hydrogen effect on oxide-semiconductor-based thin film transistors, and their corresponding application in flat panel displays and optical detecting. The present special issue on "Thin Film Transistor" can be considered as a status report reviewing the progress that has been made recently on thin film transistor technology. These papers can provide the readers with more research information and corresponding application potential about Thin Film Transistors.

Keywords: thin film transistor; OTFT; surface treated; hydrogen effect; flat panel displays; optical detecting

---

Thin film transistor (TFT) is a type of field effect transistors whose active layer, the current-carrying layer, is a thin film made by depositing an active semiconductor layer, as well as the dielectric layer over a non-conducting substrate. In general, TFTs are made at low material cost and processed at low temperature. Moreover, fabrication time of TFT is shorter than that of traditional MOSFETs. TFTs have been demonstrated in a wide variety of applications, such as active-matrix liquid-crystal displays (LCD), active-matrix organic light-emitting displays, photodetecting devices, and biosensors. A display comprises a matrix of *the* smallest addressable element called pixels. An image on the display is formed by thousands (millions) of these pixels arranged in rows and columns. The TFTs act as switches which allow the pixels to turn on and off very easily. Another promising area of application is the biosensing system where TFT arrays perform electrical excitation and sensing for the study of biological cells. In TFT biosensor, a sensitive layer converts biochemical signals into electrically measurable physical parameters.

Organic thin film transistors (OTFTs) employ organic material as the semiconductor layer, and are a promising alternative to traditional mainstream inorganic TFT. The OTFTs are well suited with flexible plastic substrates at low temperatures and equivalent with inorganic TFT in basic operation and design. The effect of device geometry on performance parameter of OTFT is examined by Arnal et al. [1]. In their work, the same material and fabrication procedure is applied to two different transistor geometries—corbino and interdigitated. Corbino geometry has less parasitic capacitance as they provide less overlapping area in comparison to the interdigitated shape. The organic inverter circuit produced by using two OTFT shows maximum gain and lowest hysteresis with a Corbino structure as the driver and an interdigitated structure as the load. Krammer et al. [2] proposed a two-step fitting method to analyze transistor model with the experimental current-voltage characteristic of OTFT. First, the output and transfer characteristic of OTFT is fitted to transistor model and the parameters of the transistor are extracted. In the second step, the dependency of extracted parameters and the channel length is determined. This two-step method gives better results by comparing calculated and measured output characteristics but fails to verify channel-length-dependence of the transistor parameters. Chang, J.-F., et al. [3] examined the effect of microstructural and molecular growth of α,ω-diperfluorohexylquaterthiophene (DFH-4T) film on charge transport. The field-effect mobility increases significantly with the deployment of DFH-4T an n-type material in organic field-effect transistors. With the increase of DFT-4T thickness from 8 to 80 nm, mobility increases from 0.01 to 1 $cm^2.V^{-1}.s^{-1}$, and the threshold voltage first decreases then increases.

The surface treatment is an important method used to remove contaminants and improve the performance of the semiconductor device. Horng, R. -H., et al. [4] demonstrated a chemical solution treatment on the n-GaN surface before Al_2O_3 was grown on metal-oxide-semiconductor (MOS); capacitor consists of an Al_2O_3/n-GaN/AlN buffer/Si substrate. The chemical treatments containing oxygen plasma, BCl_3 plasma, dilute acidic/ alkali solvents, and hydrofluoric acid were used to improve the surface quality and reduce interface state trap density. Several works have reported the surface treatment of the dielectric layer to improve the TFT performance. Most of the charge carriers are confined to the semiconductor film adjacent to the semiconductor/dielectric interface. The charge carrier transport is affected by characteristics of the dielectric surface. The surface treatment of dielectric significantly improves the majority carrier mobility and reduces the interface trap.

Wide bandgap oxide semiconductor has been widely investigated due to their high mobility and low leakage currents. Among oxide semiconductor devices, amorphous indium–gallium–zinc–oxide (α-IGZO) thin film transistors (TFTs) are strong candidates for next-generation flat-panel displays and active-matrix organic light-emitting diode (AMOLED) display applications. However, various defects exist between metal and oxygen, which affects the performance of the device. In addition, the electrical characteristics of α-IGZO TFT are affected by the presence of hydrogen impurity. Hydrogen generates free carrier in the oxide semiconductor and also acts as defect passivation. Noh, H. -Y., et al. [5] investigated the effect of hydrogen injection on the α-IGZO system at different oxygen environments. Depending on the oxygen environments, hydrogen species reduces or increases defect states. The physical mechanism of optical devices based on transparent conductive oxide (TCO) such as In-Ga-Zn-O, In-Zn-O was reviewed by Jang, J., et al. [6]. By introducing oxygen defects into TCO, an optical memory action called persistent photoconductivity (PPC) is eluded. In advanced neuromorphic system, a TCO based device with PPC phenomenon could be used as optical synaptic TFT. Overall, TCO could play an important role in transparent display and low-cost optoelectronic application.

The TFT-LCDs play a prominent role in high-resolution flat panel displays. The drain-source contact and pixel electrode layer of TFT controls the light switching function and the frame rate of the LCD. An automatic optical inspection (AOI) system performs a quality check of these layers by employing a light source. The performance of three light sources metal-halide lamps, quartz-halogen lamps, and LEDs for scanning pixel electrode of TFT was investigated by Tzu, F. -M., et al. [7]. From the result, it was found that LED is a better choice in terms of cost and performance as compared to the quartz-halogen lamp and metal-halide lamp. Due to better spatial resolution and spectrum compatibility, LED should be adopted in AOI by considering energy efficiency and performance. The defect in LCD can be detected by the human eye as it can perceive green light of spectrum 555 nm. Tzu, F. -M., et al. [8] demonstrated machine vision associated with transmission chromaticity spectrometer to detect non-uniformity of the green emission layer for TFT-LCD. The just noticeable difference was utilized as the detection criterion, according to that of the International Commission on Illumination. The presented method is an alternative to manual optical inspection as it quantifies the defects more accurately.

The present Special Issue on "Thin Film Transistor" can be considered as a status report reviewing the progress that has been made recently on thin film transistor technology.

References

1. Arnal, A.; Martínez-Domingo, C.; Ogier, S.; Terés, L.; Ramon, E. Monotype Organic Dual Threshold Voltage Using Different OTFT Geometries. *Crystals* **2019**, *9*, 333. [CrossRef]
2. Krammer, M.; Borchert, J.W.; Petritz, A.; Karner-Petritz, E.; Schider, G.; Stadlober, B.; Klauk, H.; Zojer, K. Critical Evaluation of Organic Thin-Film Transistor Models. *Crystals* **2019**, *9*, 85. [CrossRef]

3. Chang, J.-F.; Shie, H.-S.; Yang, Y.-W.; Wang, C.-H. Study on Correlation between Structural and Electronic Properties of Fluorinated Oligothiophenes Transistors by Controlling Film Thickness. *Crystals* **2019**, *9*, 144. [CrossRef]
4. Horng, R.-H.; Tseng, M.-C.; Wuu, D.-S. Surface Treatments on the Characteristics of Metal–Oxide Semiconductor Capacitors. *Crystals* **2018**, *9*, 1. [CrossRef]
5. Noh, H.Y.; Kim, J.; Kim, J.-S.; Lee, M.-J.; Lee, H.-J. Role of Hydrogen in Active Layer of Oxide-Semiconductor-Based Thin Film Transistors. *Crystals* **2019**, *9*, 75. [CrossRef]
6. Jang, J.; Kang, Y.; Cha, D.; Bae, J.; Lee, S. Thin-Film Optical Devices Based on Transparent Conducting Oxides: Physical Mechanisms and Applications. *Crystals* **2019**, *9*, 192. [CrossRef]
7. Tzu, F.-M.; Chou, J.-H. Effectiveness of Light Source on Detecting Thin Film Transistor. *Crystals* **2018**, *8*, 394. [CrossRef]
8. Tzu, F.-M.; Chou, J.-H. Optical Detection of Green Emission for Non-Uniformity Film in Flat Panel Displays. *Crystals* **2018**, *8*, 421. [CrossRef]

© 2019 by the author. Licensee MDPI, Basel, Switzerland. This article is an open access article distributed under the terms and conditions of the Creative Commons Attribution (CC BY) license (http://creativecommons.org/licenses/by/4.0/).

MDPI
St. Alban-Anlage 66
4052 Basel
Switzerland
Tel. +41 61 683 77 34
Fax +41 61 302 89 18
www.mdpi.com

Crystals Editorial Office
E-mail: crystals@mdpi.com
www.mdpi.com/journal/crystals

www.ingramcontent.com/pod-product-compliance
Lightning Source LLC
LaVergne TN
LVHW072000080526
838202LV00064B/6797